International Standards for Food Safety

Editors

Naomi Rees, BSc, PhD
Senior Scientific Officer
Food Contaminants Division
Joint MAFF/DH Food Safety and Standards Group
London, United Kingdom

**David Watson, PhD, DMS,
CChem, FRSc, CBiol, FIBiol**
Head of Branch
Additives and Novel Foods Division
Joint MAFF/DH Food Safety and Standards Group
London, United Kingdom

AN ASPEN PUBLICATION®
Aspen Publishers, Inc.
Gaithersburg, Maryland
2000

The author has made every effort to ensure the accuracy of the information herein. However, appropriate information sources should be consulted, especially for new or unfamiliar procedures. It is the responsibility of every practitioner to evaluate the appropriateness of a particular opinion in the context of actual clinical situations and with due considerations to new developments. The author, editors, and the publisher cannot be held responsible for any typographical or other errors found in this book.

Aspen Publishers, Inc., is not affiliated with the American Society of Parenteral and Enteral Nutrition.

Library of Congress Cataloging-in-Publication Data

International standards for food safety / editors, Naomi Rees, David Watson.
   p. cm.
Includes bibliographical references and index.
ISBN 0-8342-1768-6
1. Food adulteration and inspection—Standards. 2. Food industry and trade—Safety measures—Standards. I. Rees, Naomi. II. Watson, David.
TX537.I58 2000
363.19'26—dc21
99-087033

Copyright © 2000 by Aspen Publishers, Inc.
A Wolters Kluwer Company
www.aspenpublishers.com
All rights reserved.

Aspen Publishers, Inc., grants permission for photocopying for limited personal or internal use. This consent does not extend to other kinds of copying, such as copying for general distribution, for advertising or promotional purposes, for creating new collective works, or for resale. For information, address Aspen Publishers, Inc., Permissions Department, 200 Orchard Ridge Drive, Suite 200, Gaithersburg, Maryland 20878.
Orders: (800) 638-8437
Customer Service: (800) 234-1660

About Aspen Publishers • For more than 40 years, Aspen has been a leading professional publisher in a variety of disciplines. Aspen's vast information resources are available in both print and electronic formats. We are committed to providing the highest quality information available in the most appropriate format for our customers. Visit Aspen's Internet site for more information resources, directories, articles, and a searchable version of Aspen's full catalog, including the most recent publications: **www.aspenpublishers.com**
    Aspen Publishers, Inc. • The hallmark of quality in publishing
    Member of the worldwide Wolters Kluwer group.

Editorial Services: Timothy Sniffin
Library of Congress Catalog Card Number: 99-087033
ISBN: 0-8342-1768-6

Printed in the United States of America

1 2 3 4 5

Dedication

To my parents for the global adventure,
colleagues for the company, and
Chris, Michael, and Jasmin for everything else.—NR

Contents

Contributors xi

Preface xv

PART I   **THE IMPORTANCE OF INTERNATIONAL FOOD SAFETY** 1

Chapter 1 — International Food Standards: The Work of Codex 3
Alan W. Randell

    Food Standards and Trade 3
    The Codex Alimentarius 4
    Codex Standards: How Are They Developed and Used? 6
    Standards in Effect 7
    Science, Risk, and Precaution 7
    National and International Standards 9
    Conclusion 9

Chapter 2 — Food Safety and International Trade: The Role of the WTO and the SPS Agreement 11
Gretchen Stanton

    Introduction 11
    The WTO Framework 12
    Negotiation of the SPS Agreement 13
    WTO Dispute Settlement 18
    Functioning of the SPS Agreement 19

PART II — NATIONAL AND REGIONAL STANDARDS 27

Chapter 3 — Regulation of Food Additives, Contaminants, and Pesticides in the United States 29
Barbara J. Petersen

Introduction 29
Laws Governing Foods and Food Additives 29
The Food Additive Approval Process 33
Regulation of Contaminants and Pesticides................. 35

Chapter 4 — European Community Legislation on Limits for Additives and Contaminants in Food 43
Stuart A. Slorach

Introduction 43
Scientific Advisory Committees 44
Food Additives 44
Pesticide Residues 46
Veterinary Drug Residues 49
Other Contaminants 51
Conclusions and Future Developments 53

Chapter 5 — Development of Australia/New Zealand Standards.. 57
Simon Brooke-Taylor and Peter J. Abbott

Introduction 57
Food Regulation in Australia.............................. 57
Food Regulation in New Zealand 59
Joint ANZ Standards 59
WTO Obligations.. 65
The Use of Risk Analysis in Standard Setting 65
The Regulation of Food Additives......................... 66
The Regulation of Contaminants and Other Restricted
    Substances.. 72
Conclusion.. 76

Chapter 6 — Southern Common Market Standards 79
Maria Cecilia de Figueiredo Toledo

Introduction 79
Historical Background 80
Objectives of MERCOSUR 80
Institutional Structure.................................... 81
The Food Commission 84
FAO Technical Assistance 94
Conclusion.. 95

PART III SCIENTIFIC AND OTHER LEGITIMATE FACTORS 101

Chapter 7 — Exposure Assessment Supporting International Developments 103
Naomi Rees

Exposure Assessment as Part of Risk Assessment 103
National Dietary Exposure Assessments 105
International Dietary Exposure Assessments 106
Exposure Consultation 108
Function of International Exposure Assessments 112
Factors Influencing the Development of International
    Assessment Methodology 112
Conclusion 113

Chapter 8 — Methods for Estimating Dietary Exposure and Quantifying Variability 115
Barbara J. Petersen

Introduction 115
Overview of Dietary Methods and Data Available from
    Different Countries 115
Defining the Intake Scenarios 119
Dietary Intake 121
Limitations and Need for Research 131
Conclusion 132

Chapter 9 — The Need for Developing Countries To Improve National Infrastructure To Contribute to International Standards 137
Lillian T. Marovatsanga

Introduction 137
Integrated Approach to Standardization 139
National Food Control: An Outline of the Major
    Constraints 140
Areas Needing Immediate Attention To Strengthen
    National Food Control Infrastructure 140
Evaluation and Assessment of Food Control and Safety
    Programs 146
Summary 146

Chapter 10—A View from Consumers 149
Diane McCrea

Introduction 149
The Impact of Codex 149
Membership of Codex 154

Principles for Decision Making 155
Conclusion 166

PART IV   **CODEX AND OTHER INTERNATIONAL STANDARDS** **169**

Chapter 11—The Codex General Standard for Food Additives— A Work in Progress **171**
Dennis Keefe, Paul Kuznesof, Susan Carberry, and Alan Rulis

Scope and Purpose of the General Standard on
   Food Additives 171
Vertical and Horizontal Standards 172
Construction of the GSFA 172
Chronology of Developing the GSFA 174
Current Status of the GSFA 190
Future Work 191

Chapter 12—Development of the Codex Standard for Contaminants and Toxins in Food **195**
Torsten Berg

Introduction 195
Development of the GSCTF 197
Scope, Purpose, and Principles of the General Standard on
   Contaminants and Toxins in Food 204
The Content of the General Standard—Status and Future 208

Chapter 13—Codex Standards for Pesticide Residues **213**
Gerald G. Moy and John R. Wessel

Introduction 213
Need for International Standards for Pesticide Residues 214
Role of International Organizations and the Codex
   Alimentarius Commission 214
Selection of Pesticides for Elaboration of Codex MRLs 215
Joint FAO/WHO Meeting on Pesticide Residues 217
Safety of Pesticide Residues 218
Good Agricultural Practice 221
Elaboration of Codex MRLs 221
Codex Extraneous Residue Limits 222
International Regulatory Practices Involving
   Pesticide Residues 223

Chapter 14—Codex Standards for Veterinary Drug Residues **227**
Palarp Sinhaseni and Richard J. Dawson

Introduction 227
General Procedures 228

Elaboration of CCRVDF Standards and Maximum
  Residue Limits................................... 228
Joint FAO/WHO Expert Committee on Food Additives 229
Selection of Analytical Methods and Sampling for the
  Control of Veterinary Drug Residues in Foods 232
Conclusion... 232

Chapter 15—Development of Radiological Standards 235
Stuart Conney and David Webbe-Wood

Introduction 235
Radioactivity and Radiation Protection 236
International Commission on Radiological Protection 238
United Nations Scientific Committee on the Effects of
  Atomic Radiation 240
International Atomic Energy Agency 240
European Atomic Energy Community/European
  Commission 242
Limitations of International Bodies 244
Accident Regulations................................ 245
The National Radiological Protection Board 247
Assessment of Doses from Consumption of Radionuclides
  in Foods... 249
Enforcement in the United Kingdom.................... 251
Conclusion... 251

**Chapter 16—Establishment of Codex Microbiological Criteria
  for Foods** 257
Mike van Schothorst and Anthony Baird-Parker

Introduction 257
History 258
Establishment of Microbiological Criteria 259
Purposes and Applications of Microbiological Criteria 264
Conclusion... 267

Index 269

CONTRIBUTORS

Peter J. Abbott, MSc, PhD
Principal Toxicologist
Food Product Standards Program
Australia New Zealand Food Authority
Canberra, Australia

Anthony Baird-Parker, PhD
Microbiologist (retired)
Consultant Food Microbiology
Northants, United Kingdom

Torsten Berg, MSc, PhD, DESS
Head of Division
Ministry of Food, Agriculture and Fisheries
Danish Veterinary and Food Administration
Søborg, Denmark

Simon Brooke-Taylor, PhD
Program Director, Food Product Standards
Australia New Zealand Food Authority
Canberra, Australia

Susan Carberry, PhD
Review Chemist
Division of Product Manufacture and Use
Office of Premarket Approval
Center for Food Safety and Applied Nutrition
U.S. Food and Drug Administration
Washington, D.C.

Stuart Conney, BA
Principal Scientific Officer
Radiological Safety Unit
Food Standards Agency
London, United Kingdom

Richard J. Dawson
Regional Office for Asia and the Pacific (FAO/RAP)
Food and Agriculture Organization
Bangkok, Thailand

Dennis Keefe, PhD
Manager, International Affairs
Office of Premarket Approval
Center for Food Safety and Applied
    Nutrition
U.S. Food and Drug Administration
Washington, D.C.

Paul Kuznesof, PhD
Leader, Chemistry and Exposure
    Assessment Team
Division of Product Manufacture and
    Use
Office of Premarket Approval
Center for Food Safety and Applied
    Nutrition
U.S. Food and Drug Administration
Washington, D.C.

Diane McCrea, MSc
International Consultant
Food Consumer Affairs and Codex
    Alimentarius Issues
Diane McCrea Consultancy
London, United Kingdom

Lillian T. Marovatsanga, MSc, PhD
Past President (deceased)
International Union of Food Science
    and Technology
Chairwoman, Emeritus
Institute of Food Nutrition and Family
    Sciences
University of Zimbabwe
Mount Pleasant, Zimbabwe

Gerald G. Moy, PhD
GEMS/Food Coordinator
Food Safety Programme
World Health Organization
Geneva, Switzerland

Barbara J. Petersen, PhD
President
Novigen Sciences Inc.
Washington, D.C.

Alan W. Randell, MSc, PhD
Senior Food Standards Officer
Joint FAO/WHO Food Standards
    Programme
Food and Agriculture Organization of
    the United Nations
Rome, Italy

Naomi Rees, PhD
Senior Scientific Officer
Food Contaminants Division
Joint MAFF/DH Food Safety and
    Standards Group
London, United Kingdom

Alan Rulis, PhD
Director
Office of Premarket Approval
Center for Food Safety and Applied
    Nutrition
U.S. Food and Drug Administration
Washington, D.C.

Mike van Schothorst, PhD
Professor
Wageningen University
The Netherlands
Vice President
Food Safety Affairs
Nestlé, Vevey, Switzerland

Palarp Sinhaseni, PhD
Associate Professor
Department of Pharmacology
Chulalongkorn University
Bangkok, Thailand

Stuart A. Slorach, PhD, Docent
Deputy Director-General
National Food Administration
Uppsala, Sweden

Gretchen Stanton, MSc
Senior Counselor
Agriculture Division
World Trade Organization
Geneva, Switzerland

Maria Cecilia de Figueiredo Toledo, PhD
Professor of Food Toxicology
Faculty of Food Engineering
State University of Campinas-Unicamp
Campinas, SP, Brazil

David Webbe-Wood, BSc
Senior Scientific Officer
Radiological Safety Unit
Food Standards Agency
London, United Kingdom

John R. Wessel
Vice President
Health and Environment International, Ltd.
Columbia, Maryland

PREFACE

We believe that this is the first book to gather information and views about international control of food safety from around the world. Achieving this has been a pleasure, a challenge, and we believe, a necessity. Demands for safe food, against a background of increasing trade, are making international controls on food safety essential. Agreements on how to control the safety of food are now in place among the members of major trading blocks, including the European Union, the United States, Australia, New Zealand, and some South American countries. Such international developments are an important step along the road to the global harmonization of food standards. Progress along this road should be helped by the development of systematic methods of reviewing risk from food. Similarly there is a growing input to the debate on how to set international food standards from consumer groups and others, including economists.

Much of the effort to achieve global standards for food safety has been done under the auspices of the Codex Alimentarius Commission. Their extensive work over the last few decades has grown in importance with the signing of the trade and sanitary measures in the General Agreement on Tariffs and Trade (GATT). These agreements make Codex controls binding for the first time, with the World Trade Organization appointed as arbiter. Codex activity on food safety includes many subjects: food additives, residues in food from pesticides and veterinary drugs, other chemical contaminants, and toxic bacteria. International control of radiological contamination provides an interesting contrast. Such control began before the work of the Codex Alimentarius Commission and has been conducted by different world organizations.

As the potent mixture of increasing trade in food and its raw material and heightened consumer interest in food safety spreads around the world, so does the politics of food. We must emphasize, therefore, that the contributors to this book are expressing their own views, not those of the organizations that employ them or with which they are affiliated.

We hope that you find this book interesting and stimulating.

Naomi Rees
David Watson

PART I

The Importance of International Food Safety

CHAPTER 1

International Food Standards: The Work of Codex

Alan W. Randell

FOOD STANDARDS AND TRADE

Food standards for consumer protection and trade have a long tradition. In ancient Athens, beer and wine were inspected to ensure their purity and soundness. The Romans provided a well-organized food control system to protect consumers from being defrauded or sold bad quality or adulterated produce. During the European Middle Ages, various countries passed laws concerning the quality and safety of eggs, sausages, cheese, beer, wine, and bread. Some of these ancient statutes are still in operation today. The industrial revolution, which began in Europe in the late eighteenth century, gave technological and economic impetus to trade in foods across and between continents. Food chemistry and microbiology date from the late 1800s, and most modern food standards and food control systems also date from this period (FAO/WHO, 1976).

Food laws are enacted to protect consumers against unsafe products, adulteration, and fraud, as well as to protect the honest food producer and trader. They also facilitate the movement of goods within and between countries by providing a common lexicon for food quality and safety. Food laws have been traditionally considered to be both a public good and responsibility.

The spontaneous and independent development of food laws and standards by different countries had the inevitable effect of creating different sets of standards and, thus, barriers to food trade. This problem was recognized in the early years of the twentieth century, and trade associations began to form to put pressure on governments to harmonize their food standards in order to allow trade in safe foods of defined quality. One of the earliest such associations was the International Dairy Federation, founded in 1903, whose work on standards for milk and milk products was later to be an important catalyst in the development of the Codex Alimentarius Commission (CAC).

Following the creation of the Food and Agriculture Organization (FAO) in 1945 and of the World Health Organization (WHO) in 1948, these two organizations began a series of joint expert meetings on nutrition and related areas. In 1950, experts at the first meeting of the Joint FAO/WHO Expert Committee on Nutrition stated (FAO/WHO, 1950), "Food regulations in different countries are often conflicting and contradictory. Legislation governing preservation, nomenclature, and acceptable food standards often varies widely from country to country. New legislation not based on scientific knowledge is often introduced, and little account may be taken of nutritional principles in formulating regulations." Noting that the conflicting nature of food regulations may be an obstacle to trade and, therefore, affect the distribution of nutritionally valuable food, the committee suggested that FAO and WHO study these problems more closely (FAO/WHO, 1950).

One of the most critical problems to emerge from the studies of FAO and WHO was that of the use of food additives. The fourth report of the Joint FAO/WHO Expert Committee on Nutrition noted, "The increasing, and sometimes insufficiently controlled, use of food additives, has become a matter of public and administrative concern"(FAO/WHO, 1955). It also noted that the means of solving problems related to food additives may differ from country to country and stated that this fact "must in itself occasion concern, since the existence of widely differing control measures may well form an undesirable deterrent to international trade" (FAO/WHO, 1955). In the same year as the publication of this report, 1955, FAO and WHO convened the first Joint FAO/WHO Conference on Food Additives. The Joint FAO/WHO Expert Committee on Food Additives (JECFA) began work immediately and still meets regularly. At its first meeting, it enunciated the *General Principles Governing the Use of Food Additives*, a text that still forms the framework for consideration of food additive use (see FAO/WHO, 1957; FAO/WHO, 1999).

THE CODEX ALIMENTARIUS

The CAC was established at the FAO Conference in 1961 with the objective of establishing international standards to facilitate food and agricultural trade (Davies, 1970; Randell, 1995). FAO was fulfilling the mandate given to it at its founding during the United Nations Conference on Food and Agriculture held in Hot Springs, Virginia, in 1943, to establish quality standards that would enable international trade to meet the needs of the hungry in the postwar world. Recognizing that the safety of foods is an essential component of quality, FAO called on WHO to join it in this important work. In the following year, 1962, the Joint FAO/WHO Food Standards Programme was created, with the CAC as its executive organ. The CAC, an intergovernmental body open to all member countries of FAO and WHO, is an organization in which governments make decisions, but where members of civil society can provide input and comment on the draft standards. There were 165 members of the CAC as of December 1999.

In the period following the establishment of the CAC, a significant number of international standards for food commodities were developed. The CAC also gave high priority to the development of a general standard for food labeling and an international code of practice on food hygiene. Standards for the use of food additives or the presence of other chemicals in foods arising either from agricultural use or environmental contamination tended, however, to be developed on an ad hoc basis (notwithstanding the considerable number of safety evaluations undertaken by the independent FAO/WHO expert panels on food additives, contaminants, and pesticide residues).

A comprehensive approach to establishing international standards for food additives began in 1989, following the preparation of a seminal paper by Dr. W.H.B. Denner (Denner, 1989). He stressed that the regulation of food additives needed to be approached on a general basis, taking into account three major factors:

1. the safety evaluation of the substance
2. its practical and technological use in the food
3. the potential exposure of the population

Denner's paper led to the development of the Codex General Standard for the Use of Food Additives, a work that is still in progress (FAO/WHO, 1999).

At the time of the publication of Denner's paper, there had been three international FAO/WHO Conferences on Food Additives in addition to the Joint FAO/WHO Food Standards Conference of 1962. Each of these conferences set out new programs of work in the area of management of chemicals in foods, but not in an integrated manner. The FAO/WHO Conference on Food Standards, Chemicals in Foods and Food Trade held in cooperation with the General Agreement on Tariffs and Trade (GATT) (March 1991) adopted a much wider approach to the development of food standards. It stressed the importance of food quality and safety as a broad issue that affects all foods moving in international trade. In the area of food safety, it introduced the concept of risk assessment, a scientifically based process consisting of the following steps: (1) hazard identification, (2) hazard characterization, (3) exposure assessment, and (4) risk characterization. The members of that conference also adopted a recommendation that the CAC and the relevant Codex committees, responsible for the development of Codex standards, codes of practice, and guidelines, should make explicit the methods they have used to assess risk (FAO/WHO/GATT, 1991). The mutual interests of governments, industries, and consumers in the transparency and confidence of risk assessment procedures led to further recommendations calling on the CAC to describe the basis of the risk assessment methods used in arriving at its recommendations, guidelines, or standards, even though it might not prove possible to quantify all aspects of the risk assessment (FAO/WHO, 1991a). It was stressed that there was a need for uniformity of assessment, that the process should be open and available to national governments and interested organizations, and that steps should be undertaken to increase the understanding of the process by the press (FAO/WHO, 1991b).

The CAC commissioned a review of its risk assessment procedures (FAO/WHO, 1993). As a result of this review, risk assessment was included as a major program area in the Codex Program of Work. In addition, changes were proposed in the rules of procedure and in the terms of reference of relevant Codex committees.

These developments coincided with the discussions underway in the Uruguay Round of Multilateral Trade Negotiations, held under the auspices of GATT from 1986 to 1994. The Final Act of these negotiations established the World Trade Organization (WTO). Multilateral agreements, developed during the course of the Uruguay Round, recognized the importance of science-based standards in removing artificial, unjustified, or discriminatory barriers to trade in food products. Codex standards for food safety are specifically identified under the Agreement on the Application of Sanitary and Phytosanitary Standards (SPS Agreement). Codex recommendations for other aspects of food quality are used as the basis for developing national food regulations under the provisions of the Agreement on Technical Barriers to Trade (GATT, 1994).

CODEX STANDARDS: HOW ARE THEY DEVELOPED AND USED?

Codex standards, codes of practice, and guidelines are recommendations to governments that facilitate countries in accepting into their territories food that is safe and of good quality, properly labeled and packaged, and prepared under hygienic conditions. They are accompanied by recommendations on food inspection and certification systems, as well as methods of analysis and sampling to provide the framework for the application of standards to food products as they move in international trade.

The decision to develop a Codex standard or any other recommendation is taken by the CAC (or by its Executive Committee on its behalf) following an analysis of the need for the standard as a matter of consumer protection and the effect that the absence of a standard would have on international trade. Proposals to develop standards may be made by Codex member countries on the basis of action they have taken in their own legislation (e.g., registration of a new pesticide use) or made by any Codex committee.

The development of a standard to protect human health requires a risk assessment. A risk assessment is normally conducted by independent expert bodies from FAO and WHO such as the JECFA. Similarly, at the national level, many of the raw data for the safety of food additives or for agricultural and veterinary chemicals are submitted by the food or chemical industries, research institutes, and universities. The development of such data is expensive and the costs of establishing the proof of safety are, therefore, borne by the commercially interested parties, including the costs of additional research demanded by the expert committees. These parties are not, however, the exclusive source of data. Data from any source are accepted for consideration by the independent committees of FAO and WHO. All data are subjected to strict examination by the independent experts for their quality, adequacy, and integrity. Further calls for data are a common feature of the evaluation process.

Once the safety of the proposed standard has been scientifically determined, the proposal can proceed through the eight-step Codex Procedure for the Elaboration of Codex Standards and Related Texts. During the course of this procedure, the proposal is submitted twice to governments (and nongovernmental observers to Codex) for comment and review; twice to the responsible Codex committee, and twice to the CAC itself. The final adoption of a Codex standard can be made only by the plenary sessions of the CAC (FAO/WHO, 1997).

Countries use Codex standards to develop national food regulations, either directly or by reference. Sometimes the legislation makes direct reference to the applicable Codex texts, as in the case of European Union legislation on food labeling; or indicates that the Codex text has been used as the basis of the legislation, such as the European Union's Food Control Directive. In some cases, the standards are used without direct reference to their source, but the resulting regulations are fully consistent with Codex.

STANDARDS IN EFFECT

The CAC has adopted over 200 standards for individual food items; general standards in the areas of food labeling and food additives; codes of hygienic practice, both general and for specific production and handling processes; and 2,800 specific limits for the presence of chemicals that occur in foods as a result of recognized good agricultural or veterinary practices.

The different degrees of coverage of Codex standards are explained in part by the genuine problems that occur in international trade in foods. Problems such as mislabeling, nonadherence to good manufacturing practices (especially failure to adhere to requirements for high-risk canned foods), filth in foods reflecting unhygienic conditions, and the presence of pesticide residues rank high among the causes for rejections of foods moving in trade. Another reason for the difference in the nature of the coverage lies in the nature of the science that underpins Codex work. Because the understanding of chemical toxicology and risk assessment is very highly developed and accepted around the world, Codex coverage of chemical additives and residues is thorough and detailed. On the other hand, scientific understanding of microbiological problems has yet to reach the stage where risks can be fully quantified. Codex recommendations in this area are, therefore, more general and are aimed at preventing human health problems, or threats to fair practices in food trade.

SCIENCE, RISK, AND PRECAUTION

The Role of Science

In 1995, the CAC adopted four *Statements of Principle Concerning the Role of Science in the Codex Decision-Making Process and the Extent to Which Other Factors Are Taken into Account* (FAO/WHO, 1995).

1. Standards and other texts shall be based on sound scientific analysis and evidence based on a review of all relevant information to ensure the quality and safety of the food supply.
2. The CAC will have regard to other legitimate factors relevant for the health protection of consumers and the promotion of international trade.
3. Food labeling plays an important role in both of these objectives.
4. When members of the CAC agree on the necessary level of public health protection but hold differing views in regard to the other legitimate factors, they may abstain from accepting the standard without necessarily preventing a decision by Codex.

The underlying scientific evaluation processes that led to development of maximum limits for residues, additives, and contaminants in foods are supported fully by the first of these statements. However, the extent to which the CAC will take into account other legitimate factors and the question of labeling in relation to the application of these statements is a matter of ongoing debate.

Risk Analysis, Prudence, and Precaution

Since the 1991 FAO/WHO Conference on Food Standards, Chemicals in Food and Food Trade, risk analysis[1] has become accepted as the basis of decision making by the CAC. This discipline involves risk assessment, risk management,[2] and risk communication.[3] In developing Codex standards, the risk assessment component is undertaken by committees of independent experts convened by FAO and WHO, including assessments of possible exposures of human populations to chemicals in foods. The process allows decisions that admit the presence of a chemical substance in foods, whether an additive or a residue from agricultural or veterinary practices, only when there is sufficient scientific certainty to ensure that there will be no increased risk to human health.

It is recognized that so-called "zero risk" is impossible to achieve, but the approach taken by Codex and its scientific advisory bodies attempts to come as close to zero risk as possible by considering each potential threat on a case-by-case basis. The notion of "relative risk," or the balancing of one risk against another, has not been used by Codex. Care should be taken not to confuse the absence of zero risk with the presence of scientific uncertainties. The scientific evaluations used in the Codex process use conservative safety factors and prudence in arriving at conclusions within the bounds of scientific certainty to ensure that there will be no appreciable risk to human health. Under such circumstances, additional application of the precautionary principle should not be required.

The problem is more complex in the case of environmental contamination, microbiological contamination of foods, or other risks where there may be gaps in the scientific knowledge. Here the precautionary principle allows measures to be introduced to prevent or minimize threats to health until such time as better knowledge allows the application of more scientifically certain risk-based measures to be applied. This is the approach recognized in Article 5.7 of the SPS Agreement (GATT, 1994), which applies to cases where the relevant scientific evidence is insufficient.

---

[1] *Risk analysis:* A process consisting of three components: (1) risk assessment, (2) risk management, and (3) risk communication.

[2] *Risk communication:* The interactive exchange of information and opinions concerning risk among risk assessors, risk managers, consumers, and other interested parties.

[3] *Risk management:* The process of weighing policy alternatives in the light of the results of risk assessment and, if required, selecting and implementing appropriate control options, including regulatory measures.

NATIONAL AND INTERNATIONAL STANDARDS

The SPS Agreement confirms the right of WTO members to introduce their own measures to limit risks to human health. The WTO has confirmed that this is an autonomous right of sovereign states. There are, however, disciplines regarding how such measures are to be introduced and applied. Where it is not possible or it is inappropriate to base such measures on international standards, or where the international standard does not meet the level of protection established by the member country, higher or stricter standards may be introduced nationally, but these must be based on scientific principles and appropriate risk assessment. Establishing and justifying such measures raise other complex issues, such as the question of consistency in the application of levels of protection. The potential for creating new problems under such circumstances is ever present.

Generally, the level of protection inherent in Codex standards is recognized as meeting the requirements of most, if not all, of the member countries of the CAC, and the introduction of stricter standards must be considered with care in order to ensure scientific validity and to minimize negative effects on trade. Countries always have the option of convincing other members of the CAC that the higher level of protection that they wish to maintain should be reflected in Codex standards.

In general, the scientific basis of Codex standards ensures adequate health protection. By accepting Codex standards as a common basis for trade, member countries ensure these standards provide consumers, farmers, food processors, and retailers with confidence in the quality and safety of the food supply. Importantly, this is true whether the food products are of domestic origin or have been imported.

CONCLUSION

Maintaining confidence in international standards for food quality and safety requires continued input and development of new techniques and new data. The original concept of standards has been expanded from those of the early pure food acts so that those standards provide assurance to well-organized consumer movements that food is good and safe to eat. Standards also form the basis for the free movement of quality foods all over the world. In forthcoming years, the CAC will concentrate on the use of risk analysis to underpin its recommendations and standards. This will allow the recognition of equivalence of food quality and safety systems on the basis of well-defined food quality and safety objectives. The process will take into account acute risks where possible, including the assessment and management of substances known to cause allergies, as well as chronic effects. A better understanding and quantification of the wide variety of dietary patterns around the world will assist in more precisely meeting the objectives established by national governments for the protection of their populations and development of their agricultural economies.

REFERENCES

Davies, J. V. H. (1970). The Codex Alimentarius. *J Assoc Public Analysts* 8, 53–76.

Denner, W. H. B. (1989). *Future Activities of the Committee in Regard to the Establishment and Regular Review of Provisions Relating to Food Additives in Codex Standards and Possible Mechanisms for the Establishment of General Provisions for the Use of Additives in Non-Standardized Foods.* Unpublished FAO paper CX/FAC 89/6. Rome: Food and Agriculture Organization.

Food and Agriculture Organization/World Health Organization. (1950). *Joint FAO/WHO Expert Committee on Nutrition; Report of the First Session.* Technical Report Series No. 16. Geneva, Switzerland: World Health Organization.

Food and Agriculture Organization/World Health Organization. (1955). *Joint FAO/WHO Expert Committee on Nutrition; Report of the Fourth Session.* Technical Report Series No. 97. Geneva, Switzerland: World Health Organization.

Food and Agriculture Organization/World Health Organization. (1957). *General Principles Governing the Use of Food Additives: Joint FAO/WHO Expert Committee on Food Additives; Report of the First Session.* FAO Nutrition Meetings Report Series No. 15. Rome: Food and Agriculture Organization.

Food and Agriculture Organization/World Health Organization. (1976). *Guidelines for Developing an Effective National Food Control System.* Prepared in cooperation with UNEP. FAO Food and Nutrition Paper No. 1. Rome: Food and Agriculture Organization.

Food and Agriculture Organization/World Health Organization. (1991a). *Report of the Thirty-eighth Session of the Executive Committee of the Codex Alimentarius Commission.* July 1991. Unpublished FAO document ALINORM 91/4, para. 24. Rome: Food and Agriculture Organization.

Food and Agriculture Organization/World Health Organization. (1991b). *Report of the Nineteenth Session of the Codex Alimentarius Commission.* July 1–10, 1991. ALINORM 91/40, para.78. Rome: Food and Agriculture Organization.

Food and Agriculture Organization/World Health Organization. (1993). *Risk Assessment Procedures Used by the Codex Alimentarius Commission and Its Subsidiary and Advisory Bodies.* ALINORM 93/37. Rome: Food and Agriculture Organization.

Food and Agriculture Organization/World Health Organization. (1995). *Report of the Twenty-first Session of the Codex Alimentarius Commission, Rome.* July 3–8, 1995. ALINORM 95/35, paras. 28–26 and Appendix 2. Rome: Food and Agriculture Organization.

Food and Agriculture Organization/World Health Organization. (1997). Uniform procedure for the elaboration of Codex standards and related texts. In *Procedural Manual of the Codex Alimentarius Commission*, 10th ed. Rome: Food and Agriculture Organization.

Food and Agriculture Organization/World Health Organization. (1999). General standard for the use of food additives (Codex Stan 192–1995, Revised 1997). In *Codex Alimentarius*, Vol. 1A. Rome: Food and Agriculture Organization.

Food and Agriculture Organization/World Health Organization/General Agreement on Tariffs and Trade. (1991). *FAO/WHO Conference on Food Standards, Chemicals in Food and Food Trade* (in cooperation with GATT, Report). Unpublished FAO document ALINORM 91/22. Rome: Food and Agriculture Organization.

General Agreement on Tariffs and Trade. (1994). *The Results of the Uruguay Round of Multilateral Trade Negotiations; The Legal Texts.* Geneva, Switzerland: GATT Secretariat.

Randell, A. W. (1995). Codex Alimentarius: how it all began. *Food Nutr Agriculture* 13/14, 35–40.

Chapter 2

Food Safety and International Trade: The Role of the WTO and the SPS Agreement

Gretchen Stanton

INTRODUCTION

In the recent past, technological change has allowed greatly modified food production and processing, and large increases in international shipping, with reduced delivery times. Increased per capita incomes have led to increased consumer demand for food products and variety in food. This has fueled international trade in food products. Commensurate with these increases, consumer concerns about the safety of food, whether domestically produced or imported, have also increased.

International food standards can make international trade much safer and more reliable. The Codex Alimentarius Commission (CAC) was established to implement the Joint FAO/WHO Food Standards Programme—the stated purpose of which is: "protecting the health of the consumers and ensuring fair practices in the food trade ...." (CAC, 1997)

Considerable resources and expertise, both governmental and private, have been spent developing Codex standards and codes of practice to ensure that foods moving in international trade do not threaten human health. However, the use of Codex texts by governments is discretionary. With the creation of the World Trade Organization (WTO) in January 1995, the Agreement on the Application of Sanitary and Phytosanitary Measures (the SPS Agreement) also came into effect. For the first time, this agreement gives governments strong legal encouragement to use Codex standards and codes of practice because the SPS Agreement identifies them as the references against which national requirements will be measured.

THE WTO FRAMEWORK

Starting in 1945 and 1946, more than 50 countries began negotiating for a third institution to handle international economic cooperation and join the "Bretton Woods" institutions of the World Bank and the International Monetary Fund. The original ambition of the negotiators was to create an international trade organization that extended beyond world trade disciplines to include rules on employment, commodity agreements, restrictive business practices, international investment, and services. This was never realized. Instead, an interim General Agreement on Tariffs and Trade (GATT) was signed in 1948. GATT defined basic principles for international trade, with more detailed rules in certain areas, and was also an institution for the resolution of trade disputes and for the negotiation of further reductions of trade barriers. An ever-expanding membership used the GATT to conduct periodic multilateral trade negotiations. Initial negotiating rounds focused almost exclusively on the mutual reduction of import duties; later rounds began to address the more difficult issues of nontariff barriers to trade, including standards.

Trade negotiators were concerned that as tariff protections were reduced, governments might turn to other types of measures as a way to block or limit imports. Unnecessary technical requirements, for example, could severely restrict imports or give preference to products from one area over another. On the other hand, technical regulations could serve important objectives, including protection of health, improvement of quality, or limitation of practices that might deceive consumers.

The 1979 Tokyo Round of multilateral trade negotiations produced an Agreement on Technical Barriers to Trade (the 1979 TBT Agreement, also called the Standards Code), to which countries could adhere on a voluntary basis. This agreement covered all technical regulations and standards, including many sanitary regulations. Its basic principle was that whenever an international standard existed, governments should use that international standard as their requirement for imported goods. The agreement also provided governments with an open-ended list of reasons why an international standard might not be appropriate for their needs, and hence ways in which a government could justify its deviation from the standard. The 1979 TBT Agreement required governments to give advance notice of any new technical regulations to allow time for trading partners to comment. The agreement, however, never applied to all GATT members since only 48 governments accepted it. The 1979 TBT Agreement also contained dispute settlement provisions, which over time did not prove effective.

In 1986, the GATT member governments began the Uruguay Round of multilateral trade negotiations. One major objective of the negotiations was to establish more effective trade rules for agricultural products. The efforts to reduce the use of nontariff barriers to agricultural trade led to a complementary agreement on the use of sanitary measures. This agreement, based in part on the 1979 TBT Agreement, established more detailed disciplines for trade measures whose intended objective was to ensure food safety, animal health, or plant protection. The SPS Agreement was negotiated separately, but parallel to, the Agreement on Agriculture and the revised 1995 TBT Agreement. These agreements were part of the package of 28 trade agreements adopted in April 1994, when governments agreed to establish the WTO.

The members[1] of the WTO are legally committed to respect the rules contained in all the agreements.[2] All members may participate in the governing General Council and the councils for trade in goods and for trade in services, as well as in the Dispute Settlement Body. All members are also automatically members of the distinct committees that oversee the implementation of the various agreements.

NEGOTIATION OF THE SPS AGREEMENT

Concerns about the effects of sanitary measures on international trade were included among the original list of issues to be addressed when the Uruguay Round negotiations began in 1986. These concerns grew as the negotiations aiming at a fairer and more market-oriented agricultural trading system evolved, and in 1988, a specific working group on sanitary measures was created.

Throughout the negotiations, various concerns were identified by different countries. In 1985, the European Community (EC) decided to ban imports of beef from cattle treated with growth hormones. The ban was to take effect in 1988, but was subsequently delayed until 1989. In March 1987, the United States brought the issue to GATT and claimed violation of the 1979 TBT Agreement. The United States requested that a technical experts group be established according to the dispute settlement provisions of that agreement. The request was denied, because the 1979 TBT Agreement applied only to final product requirements and not to production or processing methods.[3]

At the same time, a number of Latin American countries, notably Argentina, Uruguay, and Chile, sought to use the negotiations to ensure changes to long-standing restrictions that barred their meat from the profitable markets of North America and Japan because of foot-and-mouth disease.

While the negotiations were underway, the United States banned imports of certain European wines because of residues of the fungicide procymidon, which was not registered and approved for use in the United States. This incident heightened European interests in the negotiations. Other countries with strong agricultural export interests (including Australia, Canada, and New Zealand), major import interests (Japan), or historical involvement with the Standards Code (the Nordic countries) were also very active in this attempt to clarify the proper use of trade restrictions for food safety and animal/plant health protection.

---

[1] At the end of 1999, 135 governments were members. At that date, 30 more governments were negotiating their accession to the WTO; it is expected that membership will continue to increase in coming years.

[2] With the exception of two limited-membership or plurilateral agreements, which concern government procurement and trade in civil aircraft.

[3] The EC measure did not establish final product requirements, such as a maximum residue level for the hormones in question. The EC requirement was on the production process (i.e., growth-promoting hormones could not be administered to the animals).

Basic Provisions of the SPS Agreement

The SPS Agreement attempts to conciliate food safety and international trade by recognizing the sovereign right of any government to take measures to protect life or health, while ensuring that this right is not misused for protectionist purposes to create unnecessary barriers to trade. One basic principle included in the SPS Agreement is that measures intended to protect health must have a scientific basis. Governments can restrict international trade when this is necessary to ensure safe food or animal/plant health, but the necessity must be determined on the basis of scientific evidence and analysis. Specifically, the text of the SPS Agreement indicates that governments "shall ensure" that a sanitary measure is applied "only to the extent necessary to protect human, animal or plant life or health, is based on scientific principles and is not maintained without sufficient scientific evidence" with an exception that permits provisional measures to be taken when "scientific evidence is insufficient (GATT 1994)."[4]

This basic obligation raises the question of how governments can ensure that their sanitary measures have a sufficient scientific basis. The SPS Agreement provides two options: (1) governments can make use of internationally developed standards, guidelines and recommendations, or (2) governments can base their measures on risk analysis. In fact, decisions by governments to use an international standard will normally require some analysis of the risks involved.

Basing national requirements on internationally developed standards, guidelines, and recommendations is referred to as "harmonization" in the SPS Agreement (GATT, 1994).[5] The agreement explicitly identifies three international standard-setting bodies as relevant in this context: (1) the Joint FAO/WHO CAC with respect to food safety,[6] (2) the Office International des Epizooties (OIE) for animal health and zoonoses,[7] and (3) standards developed under the auspices of the FAO's

---

[4]Article 2.2 of the SPS Agreement states in full: "Members shall ensure that any sanitary or phytosanitary measure is applied only to the extent necessary to protect human, animal or plant life or health, is based on scientific principles and is not maintained without sufficient scientific evidence, except as provided for in paragraph 7 of Article 5."
Article 5.7 of the SPS Agreement provides:
In cases where relevant scientific evidence is insufficient, a Member may provisionally adopt sanitary or phytosanitary measures on the basis of available pertinent information, including that from the relevant international organizations as well as from sanitary or phytosanitary measures applied by other Members. In such circumstances, Members shall seek to obtain the additional information necessary for a more objective assessment of risk and review the sanitary or phytosanitary measure accordingly within a reasonable period of time.

[5]The SPS Agreement makes no distinction between an international standard, guideline, or recommendation. For convenience, the term *international standard* is used in this chapter to refer to standards, guidelines, or recommendations.

[6]Annex A of the SPS Agreement defines *international standards, guidelines and recommendations* to comprise, for food safety, "the standards, guidelines and recommendations established by the Codex Alimentarius Commission relating to food additives, veterinary drug and pesticide residues, contaminants, methods of analysis and sampling, and codes and guidelines of hygienic practice." A footnote to the definition clarifies that contaminants include pesticide and veterinary drug residues and extraneous matter.

[7]Zoonoses here refers to diseases of animals that can infect humans, such as rabies.

International Plant Protection Convention (IPPC) with respect to plant health. These standards and recommendations are presumed to be based on scientific principles and supported by scientific evidence, and to not restrict trade more than is necessary to ensure protection of health. A government that applies one of these international standards as its national requirement is presumed to be complying with the SPS Agreement. A government has the right to impose a measure that is more trade restrictive than the international standard, but in that case, it can be required, in a trade dispute, to justify its measure.[8]

The SPS Agreement indicates that a more restrictive requirement may be justified either because "there is a scientific justification" (GATT, 1994) or because the international standard does not provide the level of health protection the government considers appropriate. In both cases, reference is made to the risk analysis provisions of the agreement,[9] and the distinction between the two exceptions is not very apparent. This question may eventually need to be clarified by a dispute settlement panel.[10] One possible approach is to consider the first exception ("if there is a scientific justification") to apply when a government considers that the international standard is out of date. Another example might be when a government considers that the international standard does not take into account unusual dietary patterns or climatological conditions. In other words, the government does not agree with the scientific premise of the standard. The second exception ("as a consequence of the level of . . . protection") (GATT, 1994) could be considered to apply when the government finds no fault with the scientific justification of the standard, but desires a higher level of protection than would result from use of the standard.

Reliance on international standards raises several concerns, which were extensively discussed during the negotiations. Before the negotiation of the SPS Agreement, Codex standards and recommendations and OIE guidelines were developed to promote food safety and ensure fair trade practices, but there was no expectation that they would serve as reference standards in trade disputes.[11] The question was

---

[8] In January 1998, the first disputes related to the SPS Agreement, concerning the EC prohibition on imports of beef treated with certain growth-promoting hormones, were considered by the WTO's Appellate Body. Although the report of the Appellate Body (WTO, 1998a) addresses the application of the SPS Agreement only to the case at hand, its finding could further clarify some of the basic provisions. For example, the Appellate Body indicated that the right of a WTO member to establish its own level of protection is an autonomous right, but it is not "an absolute or unqualified right . . ." (para. 173, WT/DS/26/AB).

[9] Article 3.3 indicates that governments may impose sanitary measures that result in a higher level of protection than that afforded by the international standard if "there is a scientific justification, or as a consequence of the level of sanitary or phytosanitary protection a Member determines to be appropriate in accordance with the relevant provisions of paragraphs 1 through 8 of Article 5" (GATT, 1994). A footnote to this sentence states, "For the purposes of paragraph 3 of Article 3, there is a scientific justification if, on the basis of an examination and evaluation of available scientific information in conformity with the relevant provisions of this Agreement, a Member determines that the relevant international standards, guidelines or recommendations are not sufficient to achieve its appropriate level of sanitary or phytosanitary protection."

[10] The Appellate Body report on EC hormones stated that the distinction between these two situations "may have very limited effects and may, to that extent, be more apparent than real" (WTO, 1998a).

[11] The IPPC began to establish international standards only subsequent to the entry into force of the SPS Agreement.

raised whether these standards, and particularly some of the older Codex standards, were based on the best available scientific evidence. Another concern was whether the standards and recommendations resulted in a consistent level of risk. After considerable discussion, the negotiators agreed there was no better alternative. The objective of these standard-setting bodies was to identify the measures necessary to ensure health protection. Many countries, in particular developing countries, lack the resources to develop their own, scientifically justified requirements, and the international standards represent the best internationally acceptable technical evidence. The close involvement of the three standard-setting bodies in the negotiations,[12] and the Codex decision in 1991 (CAC, 1997) to regularly review older standards, helped alleviate some of these concerns.

Another concern is the capacity of developing countries, in particular the less developed ones, to use international standards. Development of their own standards, however, is much more difficult for these countries. Moreover, where they have particular export interests, developing countries have often been able to meet the stringent (and often differing) standards imposed by their potential markets. The use of international standards (and the equivalency provisions described below) offer the developing countries the possibility of avoiding what may be arbitrarily different and costly requirements. Furthermore, developing countries may request a time-limited exception to all or part of their obligations under the SPS Agreement, including the use of international standards. To date, no country has made such a request. This is an encouraging sign that these countries understand the advantage that the use of international standards provides to them, despite the difficulties of application.[13]

A further concern was the possible effect this new role for the international standards might have on the standard-setting process itself. The negotiators worried that the process of standard adoption would come under increased political pressure to ensure that the reference standards were politically acceptable irrespective of their technical merit. This pressure could significantly impede the process of standard development at a time when more standards are needed to assist governments, and particularly less developed countries, to meet the obligations of the SPS Agreement. Although political pressures may have sometimes been evident in the past with regard to some Codex decisions, they have become more noticeable at recent sessions of the CAC, and to some extent also at the OIE.[14] The explicit recognition of Codex, OIE, and IPPC standards by the WTO has been both a blessing

---

[12]For instance, at the request of the group negotiating the SPS Agreement, the OIE reviewed and revised its recommendations regarding foot-and-mouth disease.

[13]The SPS Agreement encourages governments to provide technical assistance, either bilaterally or through the relevant international organizations, to developing countries. The WTO Secretariat has an active program of assisting developing countries to understand and make use of the SPS Agreement.

[14]It should be noted, however, that discussions of bovine spongiform encephalopathy (BSE), a politically sensitive issue, have been featured prominently at recent meetings of the OIE. (See, for example, *Final Report of the 67th General Session of the International Committee of the OIE*, May 1999. Paris: Office International des Epizooties, pp. 28–30; also *Final Report of the 66th Session of the International Committee of the OIE*, May 1998. Paris: Office International des Epizooties, pp. 52–54 and *Final Report of the 65th Session of the International Committee of the OIE*, May 1997. Paris: Office International des Epizooties, pp. 38–39.) The existence of the SPS Agreement may thus not be responsible for any increased political pressures at the OIE.

and a curse for these organizations—giving their output an important and effective role in international trade, but at the same time, exposing them to increased political tensions. These developments suggest that procedures for the adoption of international standards may have to be revised in order to guarantee that political, economic, or other concerns do not overrule the scientific evidence.

The SPS Agreement requires the SPS Committee to develop a procedure to monitor the use of international standards by WTO members. A provisional procedure was adopted by the committee in October 1997. The committee is to establish a list of those international standards that governments have identified as of particular importance in light of their trade concerns. All WTO members will subsequently be asked to indicate whether or not they base their sanitary requirements on these international standards and, if not, why not. The purpose of the exercise is to identify where important international standards are not being used because governments consider them inappropriate, or where there is a need for the development of new standards.

The equivalence provisions of the SPS Agreement may also have some relevance to the use of international standards. A government must recognize as equivalent the sanitary measures of another country when these result in a product that meets the importing country's acceptable level of risk. The agreement foresees an essentially bilateral process in which the exporting country must demonstrate that its product is equally safe, and the importing country may verify these claims. The concept of equivalence is also included in the Codex Principles of Food Import and Export Inspection and Certification and is being discussed in the corresponding Codex committee. In addition, a number of bilateral agreements recognizing the equivalency of veterinary procedures have been negotiated in recent years (FAO/WHO, 1999).

In situations where no relevant international standard, guideline, or recommendation exists, or when a government decides to not apply that standard, the SPS Agreement requires WTO members to ensure that their measures are based on an assessment of the risks. The agreement further specifies the factors that are to be considered in conducting the risk assessment.[15]

---

[15]The SPS Agreement provides as follows in Article 5:
1. Members shall ensure that their sanitary or phytosanitary measures are based on an assessment, as appropriate to the circumstances, of the risks to human, animal or plant life or health, taking into account risk assessment techniques developed by the relevant international organizations.
2. In the assessment of risks, Members shall take into account available scientific evidence; relevant processes and production methods; relevant inspection, sampling and testing methods; prevalence of specific diseases or pests; existence of pest- or disease-free areas; relevant ecological and environmental conditions; and quarantine or other treatment.
3. In assessing the risk to animal or plant life or health and determining the measure to be applied for achieving the appropriate level of sanitary or phytosanitary protection from such risk, Members shall take into account as relevant economic factors: the potential damage in terms of loss of production or sales in the event of the entry, establishment or spread of a pest or disease; the costs of control or eradication in the territory of the importing Member; and the relative cost-effectiveness of alternative approaches to limiting risks. (GATT, 1994)

Paragraph 177 of the *Report of the Appellate Body on EC Hormones* noted that the requirements of a risk assessment as well as of sufficient scientific evidence are "essential for the maintenance of the delicate and carefully negotiated balance in the SPS Agreement between the shared, but sometimes competing, interests of promoting international trade and of protecting the health of human beings (WTO, 1998a)."

Relationship with the TBT Agreement

The text of the 1995 TBT Agreement explicitly excludes any measure covered by the SPS Agreement, many of which would otherwise be considered as technical barriers to trade. However, the distinction between the measures covered by each agreement is not always easy to make. The TBT Agreement covers all technical regulations and standards (i.e., mandatory and voluntary requirements, respectively) and those measures required to assess the conformity of imported goods with the technical regulation or standard. The TBT Agreement is not limited to food or agricultural products, nor to health protection measures. It also covers, for example, industrial goods and regulations to protect consumers from deceptive practices. In the area of food trade, the TBT Agreement applies, among other things, to nutrition requirements, to labeling and packaging regulations (unless these are directly related to food safety), to quality and grade standards, and to composition requirements. In terms of protection of human health, regulations on pharmaceuticals, for example, are covered by the TBT Agreement. Although maximum residue levels for pesticides in foods are sanitary measures under the SPS Agreement, regulations to protect the health of those producing and applying the pesticides are within the scope of the TBT Agreement.

The revised 1995 TBT Agreement continues the same basic principles of the 1979 agreement. It encourages governments to make use of international standards whenever they have need of a technical regulation or standard. International standards are defined as those developed by a standard-setting body open to membership to at least all WTO governments. Codex clearly falls within this definition, even though its standards are not explicitly identified in the TBT Agreement. Codex standards of relevance to the TBT include those on food quality, composition, and labeling. The TBT Agreement, however, permits deviation from the international standards to be justified for a number of reasons, not just on the basis of scientific evidence. In this respect, it differs from the SPS Agreement.

The provisions under both the TBT and the SPS Agreements regarding the publishing of any technical regulations and the advance notice of new measures are virtually identical. Both also require governments to identify an inquiry point or office responsible for responding to all requests for information on proposed or existing technical (or sanitary) measures. The TBT Agreement also contains a code of practice for nongovernmental standard-setting bodies within WTO members.

WTO DISPUTE SETTLEMENT

A unified procedure (Figure 2–1) exists to resolve trade disputes arising from any of the agreements. If efforts to resolve trade problems through formal bilateral consultations prove unsuccessful, a member government can request that a dispute settlement panel determine whether its rights have been violated by another member.

Dispute settlement panels are normally composed of three individuals, selected following consultations with the governments involved in the dispute. Panel members are chosen for their expertise in WTO matters, and serve in a personal capacity, not as representatives of their country. A panel receives written and oral submissions from the parties to the dispute, as well as from any other interested WTO member. A panel's conclusions as to the violation (or not) of WTO obligations are presented in a report that is submitted to the WTO's Dispute Settlement Body. The panel report is automatically adopted, unless there is a consensus for nonadoption. Alternatively, the findings of a panel can be appealed before the autonomous WTO Appellate Body. Three of the standing judges from the Appellate Body review the panel's legal interpretations and procedures, and may either uphold or overturn the findings. A decision of the Appellate Body is final unless all WTO members agree by consensus to reject it.

A timetable is established for every phase of the dispute settlement process. Most panel reports are issued within nine months of establishing the panel. In disputes relating to sanitary measures, the SPS Agreement states that panels should seek scientific or technical advice as necessary. This inevitably leads to some delays in the process. Nonetheless, a government can normally expect to have a final decision on its complaint (including the Appellate Body's decision) within 12 to 15 months after requesting a panel.

Once the report of the panel (or of the Appellate Body) has been adopted, governments are given a reasonable time to change any measures found in violation of the WTO. What is reasonable varies with each situation, however; a government must regularly report on the steps it is taking to rectify the situation. If a government cannot make the necessary modifications within an acceptable period of time, it can offer to provide temporary compensation to the adversely affected member. The compensation is easier trade access for other goods (or services) if agreed to by the affected member. If a country cannot rectify the violation within a reasonable period of time, and compensation is not accepted, the adversely affected government can request authorization from the Dispute Settlement Body to take retaliatory action against the offending country. Such retaliatory action would normally take the form of higher tariffs or more restricted access for goods or services exported by the "losing" country. Although provisions for retaliation existed previously in the GATT procedures, authorization to use them has rarely been requested.

FUNCTIONING OF THE SPS AGREEMENT

Not enough time has passed to assess the value of the SPS Agreement. The SPS Agreement took effect on January 1, 1995. Except for the notification and information exchange provisions, developing countries were permitted to delay implementation of the agreement until January 1997. The least-developed countries do not have to implement the agreement until 2000.

The SPS Committee first met in March 1995 and has since met three times per year. The role of the committee is to serve as a forum for discussion of any matter related to the SPS Agreement. Specific tasks for the committee were identified in the agreement. In addition to monitoring the use of international standards, the committee is to develop guidelines to help governments avoid arbitrary differences in the levels of risk they accept, and to undertake a review of the implementation of the agreement.

20 INTERNATIONAL STANDARDS FOR FOOD SAFETY

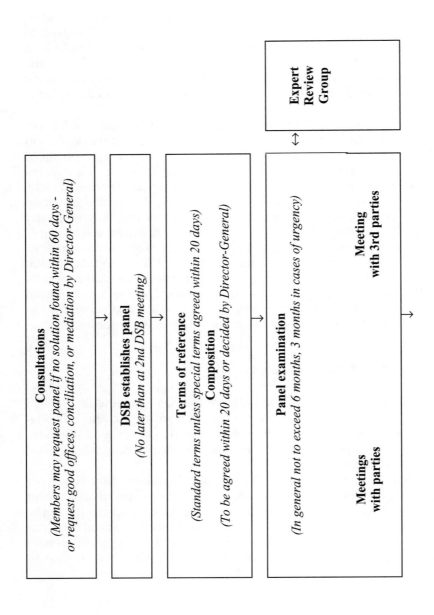

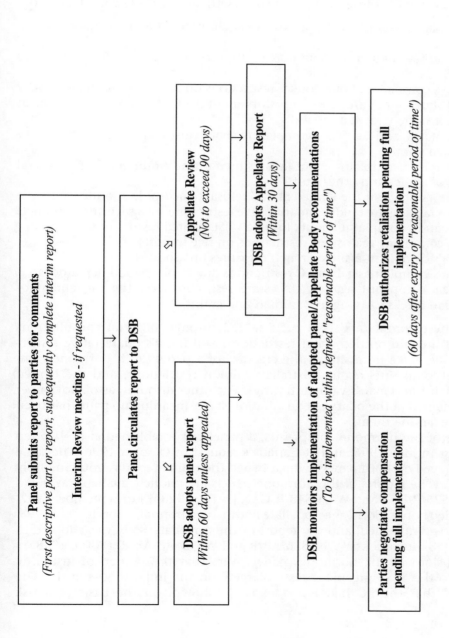

Figure 2–1 WTO Dispute Settlement Flow Chart. Courtesy of the World Trade Organization, Geneva, Switzerland.

The initial meetings focused on procedural aspects, including the procedures for the advance notification of proposed sanitary measures. The focus of SPS Committee meetings has changed as governments increasingly use the meetings as a forum to question the justification of measures imposed by trading partners. BSE-related measures, *Salmonella* regulations, and restrictions on grains are examples of the types of issues that have been on the agenda of recent committee meetings.

As of November 15, 1999, several complaints relating to the SPS Agreement had been formally brought before WTO for dispute settlement. These complaints are

1. United States' complaint about Korea's shelf-life requirements for processed meats
2. United States' complaint about Korea's inspection requirements for fruits and vegetables
3. Canada's complaint about Korea's restrictions on bottled water treatment
4. United States' and Canada's complaints about the EC ban on imports of meat from hormone-treated animals
5. United States' and Canada's complaints about Australia's ban on imports of uncooked salmon
6. United States' complaint about Japan's quarantine requirements for varietal testing of apples and other fruits
7. EC complaint about the United States' refusal to accept EC poultry
8. EC complaint against India's quantitative restrictions on agricultural products
9. Switzerland's complaint about Slovakia's BSE-related restrictions
10. India's complaint against EC restrictions on rice imports
11. Canada's complaint against French measures on asbestos
12. Canada's complaint against EC restrictions due to pine wood nematodes
13. Canada's complaint against U.S. state restrictions on the movement of Canadian trucks carrying live animals and grains

Three of these complaints have been settled through bilateral consultations (Korea's shelf-life and bottled water restrictions, and India's quantitative restrictions) while five have necessitated the creation of a panel (both hormone complaints, both complaints about Australia's salmon restrictions, and the United States' complaint on Japan's varietal testing). The other disputes, as of November 15, 1999, remained at the bilateral consultation stage, meaning no panel has been requested to examine them.

In the case of the hormone complaints, a panel was established at the United States' request in May 1996, and at Canada's request in October 1996. The same individuals served on both panels, which issued their final reports simultaneously in June 1997. (The reports of the hormone panels are publicly available as documents WT/DS26/R/USA and WT/DS48/R/CAN [WTO, 1998b].) In September 1997, the EC appealed both reports; the Appellate Body issued its final report in June 1998. The decision was appealed, and the report of the Appellate Body was adopted in November 1998 (documents WT/DS26/AB/R and WT/DS48/AB/R) (WTO, 1998a).

The conclusions of the hormones panels were that the burden of justifying conformity to the SPS Agreement rested largely on the importing country. The panels found that the EC had not adequately shown, on the basis of a risk assessment, that the Codex standards (for three of the six hormones at issue) were not sufficient to ensure the EC's desired level of health protection. The panels further concluded that the EC's prohibition on imports of hormone-treated beef could not be justified given that the EC permitted the same hormones to be used for therapeutic purposes.

The Appellate Body upheld some of the panels' findings, and overturned others. The Appellate Body clarified that a complaining government must establish a prima facie case of violation of WTO provisions; then the burden of proof shifts to the defending government. The Appellate Body also concluded that Article 3.3, which allows a government to not apply an international standard subject to certain conditions, was not an exception to the basic obligations of the SPS Agreement, but an autonomous, albeit qualified, right (see note 9 of Article 3.3 [WTO, 1998a]). The Appellate Body upheld the findings of the panels that the EC import ban was not based on a risk assessment, stating that the results of the risk assessment must sufficiently warrant and reasonably support the measure taken. At the same time, the Appellate Body noted that the risk assessment need not be quantitative, and could set out both mainstream as well as divergent views, but the risks assessed needed to be ascertainable. In this the Appellate Body agreed with the panels that theoretical uncertainties were not the kind of risk to be assessed.

The Appellate Body disagreed with the panels' findings that the EC had acted inconsistently by accepting no risk from added hormones, while not limiting exposure to the same hormones naturally occurring in greater quantities in other foods or when used for therapeutic purposes. The Appellate Body, which does not have the opportunity to seek expert advice, considered that there was "a fundamental distinction between added hormones (natural or synthetic) and naturally-occurring hormones in meat and other foods" (WTO, 1998a). On the other hand, the Appellate Body agreed with the panels that the distinction in treatment by the EC of the hormones at issue and other veterinary drugs was "arbitrary and unjustified," but did not uphold the panels' findings that this different treatment "resulted in discrimination or a disguised restriction" to trade (WTO, 1998a).

The report of the Appellate Body on EC hormones was adopted by the Dispute Settlement Body on February 13, 1998 (WTO, 1998a). Following arbitration, the EC was given until May 13, 1999, to comply with the decisions. When the EC indicated that it would not be able to comply by that date, both the United States and Canada requested the right to retaliate against EC products to compensate for their lost trade opportunities. The original panel was reconvened to determine what would be the appropriate amount of retaliation. Its final decision was issued on July 12, 1999. Since the end of that month, both the United States and Canada have been imposing 100 percent tariffs on various products imported from EC member countries.

Although the salmon and varietal testing cases do not directly deal with food safety, some of the legal findings could nonetheless be of interest in this context. In the salmon case, the panel and Appellate Body examined the assessment done by Australia of the risks of introducing fish diseases through salmon imports for human consumption. They found the risk assessment to be lacking in several critical aspects. In addition, the difference in levels of protection sought by Australia from the risk of salmon diseases compared to some other risks accepted by the country was found to be unjustifiable and a disguised restriction of trade (WTO, 1998c).

An interesting issue in the varietal testing dispute concerns Article 5.7, which in cases where sufficient scientific evidence is not available allows countries to take provisional measures. The panel and Appellate Body ruled that countries maintaining provisional measures on the basis of Article 5.7 have to actively search for scientific evidence that could enable them to undertake a risk assessment to justify their measure (WTO, 1999).

The use of the WTO dispute settlement procedures to resolve trade disputes involving sanitary measures is one indication of the effectiveness of the SPS Agreement. It is encouraging that a number of trade problems have been resolved at the bilateral level, without recourse to a dispute settlement panel. Governments, knowing that their requirements cannot be justified under the SPS Agreement, often make the necessary modifications in order to avoid a formal challenge.

In 1998, the SPS Committee will review the operation of the agreement.[16] Changes to the substantive provisions of the agreement are unlikely, given that a consensus is required for any decision by the SPS Committee. Some governments believe the agreement is too weak, while others consider it too constraining. The SPS Committee may, however, agree to improve some procedural aspects of the agreement, including those related to notifications and information exchange.

Since 1995, the SPS Agreement has imposed an enforceable set of rules on the use of food safety as well as animal and plant health regulations, to ensure that they are used for health protection purposes and not for trade protectionism. It will require several years, and the development of a body of case law, before the effectiveness of this new agreement, with its reliance on science and risk analysis, is known.

REFERENCES

Codex Alimentarius Commission. (1997). Statutes of the Codex Alimentarius Commission, Article 1(a). In *Codex Alimentarius Commission Procedural Manual*, 10th ed. Rome: Codex Alimentarius Commission, p. 4.

General Agreement on Tariffs and Trade. (1994). The Agreement on the Application of Sanitary and Phytosanitary Measures. In *The Results of the Uruguay Round of Multilateral Trade Negotiations: The Legal Texts*. Geneva, Switzerland: General Agreement on Tariffs and Trade, pp. 69–84.

Food and Agriculture Organization/World Health Organization. (1999). ALINORM 99/30, Appendix I and ALINORM 99/37. *Report of the Twenty-third Session of the Codex Alimentarius Commission*, July 1999. Rome: Food and Agriculture Organization/World Health Organization, paras. 172–175 and para. 217.

---

[16]Article 12.7 of the Agreement states: "The Committee shall review the operation and implementation of this Agreement three years after the date of entry into force of the WTO Agreement, and thereafter as the need arises. Where appropriate, the Committee may submit to the Council for Trade in Goods proposals to amend the text of this Agreement having regard, inter alia, to the experience gained in its implementation."

World Trade Organization. (1998a). *Report of the Appellate Body on EC Measures Concerning Meat and Meat Products (Hormones)*. WT/DS26/AB/R and WT/DS48/AB/R, adopted 13 February 1998. Geneva, Switzerland: World Trade Organization.

World Trade Organization. (1998b). EC Measures Concerning Meat and Meat Products (Hormones)—Complaint by the United States: Report of the Panel. WT/DS26/R/USA. Adopted on 13 February 1998; and EC Measures Concerning Meat and Meat Products (Hormones)—Complaint by Canada: Report of the Panel. WT/DS48/R/CAN. Adopted on 13 February 1998. Both reports: Geneva, Switzerland: World Trade Organization.

World Trade Organization. (1998c). Australia—Measures Affecting Importation of Salmon. Report of the Appellate Body—Complaint by Canada. WT/DS18/AB/R. Adopted on 20 October 1998. Geneva, Switzerland: World Trade Organization.

World Trade Organization. (1999). Japan—Measures Affecting Agricultural Products. Report of the Panel. WT/DS76/R; Japan—Measures Affecting Agricultural Products. Report of the Appellate Body. WT/DS76/AB/R. Both adopted on 19 March 1999. Geneva, Switzerland: World Trade Organization.

PART II

National and Regional Standards

CHAPTER 3

Regulation of Food Additives, Contaminants, and Pesticides in the United States

Barbara J. Petersen

INTRODUCTION

Foods and components in those foods are regulated by a variety of different governmental bodies in the United States, including local municipalities, state governments, and the federal government. It is a complex and sometimes overlapping system of laws and regulations. Food traveling via interstate commerce or being imported into the United States is generally regulated under federal statutes. The federal regulations governing food additives and other substances are available in Title 21 of the *U.S. Code of Federal Regulations* (21 CFR). Many states also have their own food laws. Typically when federal and state laws conflict, the doctrine of preemption will be applied. That is, if a federal standard applies, it will preempt any less restrictive state or local standard. If a state or local standard is stricter or more comprehensive than a corresponding federal standard, usually the state or local standard will apply, absent a federal determination that there is specific statutory language from Congress providing for federal preemption. An example of a state law having authority over federal law is the California Safe Drinking Water and Toxic Enforcement Act of 1986 (generally called Proposition 65) (California Health & Safety § 25249.5–25249.12 [1986]). Proposition 65 is more restrictive than most federal regulations, particularly regarding the consumer right-to-know provisions.

The purpose of this chapter is to discuss specifically those regulations that apply to food additives and contaminants. Many food additives are regulated under the same statutes and regulations that are applied to food. Where appropriate, both are discussed.

LAWS GOVERNING FOODS AND FOOD ADDITIVES

The main federal law governing food additives is the U.S. Federal Food, Drug and Cosmetic Act (FFDCA, 1938). FFDCA was initially enacted in 1938. It has been amended many times. The two most recent amendments were the Food Quality Protection Act (FQPA), passed in August 1996, and the Food and Drug Administration Modernization Act (FDAMA), passed in 1997. Although there are many components to FFDCA, in its simplest summary, FFDCA prohibits the adulteration or misbranding of any food item under the jurisdiction of the FFDCA (e.g., food and food additives moving in interstate commerce) (Food and Drug Law Institute, 1996).

The CFR is a codification of the rules published in the *Federal Register*. It encompasses all federal regulations in 50 volumes. The food additive regulations are contained in Title 21 of the CFR, while the environmental regulations are contained mainly in Title 40. Each volume of the CFR is revised once each calendar year and updated on a quarterly basis. Another useful document, the U.S. FDA's *Compliance Policy Guide* is accessible via the Internet (http://www.fda.gov) (FDA, 1998a).

The definition of *food* includes articles used for food or drink by man or other animals as well as chewing gum. Components of food including food additives are also defined as food. FFDCA also includes within the definition of food substances migrating to food from food contact articles. FFDCA also defines food additives and regulates those substances under the general provisions governing food items and under specific food additive provisions.

FQPA of 1996 amended FFDCA and established new regulations concerning pesticide residues in raw and processed food. FQPA repealed the infamous Delaney Clause with respect to pesticide residues and established a single health-based standard for pesticides regardless of whether the food was raw or processed. FQPA left in place the Delaney Clause as it applies to foods, food ingredients, or food additives. Thus, no food additive is permitted if it has been shown to cause cancer in man or animals (see below). FQPA directs the Environmental Protection Agency (EPA) to consider children's exposures and risks, establishes screening for potential estrogenic effects of pesticides, and requires exposure assessments to include all sources of similarly acting compounds. The specific details of FQPA are provided in McKenna & Cuneo's (1996) in-depth discussion of its provisions.

The Federal Meat Inspection Act (FMIA) and the Poultry Products Inspection Act authorized the U.S. Department of Agriculture (USDA) to regulate meat and poultry and meat and poultry products (Federal Meat Inspection Act, 1967; Poultry Products Inspection Act, 1957). Although *meat* is not defined in the FMIA, by common understanding it refers to the edible flesh of mammals. USDA also regulates eggs and egg products under the Egg Products Inspection Act (Egg Products Inspection Act, 1970). All plants processing meat, poultry, and eggs must have a USDA inspector present during processing. All remaining products are under FDA's regulation. FDA does not have inspectors in plants on a continuing basis, but rather makes periodic inspections. In practice, most nonmeat, poultry, and egg facilities are inspected more frequently by state inspectors than by FDA.

In 1996, USDA's Food Safety and Inspection System published new regulations requiring processors to incorporate the principles of hazard analysis critical control point (HACCP) into their standard procedures (Food and Safety Inspection Service, 1997). HACCP plans are designed to identify the critical control points for each process and to establish critical limits that must be controlled. HACCP plans must include monitoring requirements, corrective action plans, extensive recordkeeping, and verification procedures.

Labeling

USDA regulates labeling of meat and poultry products, while FDA regulates all other food labeling. The procedures used by the two agencies are quite different. USDA requires labels to be preapproved; FDA does not. Generally, the requirements are similar. The Nutrition Labeling and Education Act (NLEA), passed in 1990, spelled out extensive rules for labeling of foods (NLEA, 1990). The act was passed to provide consumers with additional information about their diets. FDA issued implementing regulations in 1993 and 1994, which further defined the requirements for nutrition labeling (21 CFR Part 101 [1994]). Expressed or implied nutrient content claims are prohibited unless they comply with the FDA nutrition-labeling regulations. Health claims are allowed, including claims for some specified relationships between a nutrient or a food and the risk of a disease or health-related conditions. Permitted claims are

1. calcium—osteoporosis
2. fat—cancer
3. saturated fat and cholesterol—coronary heart disease (CHD)
4. fiber-containing grain products, fruits, and vegetables—cancer
5. fruits, vegetables, and grain products that contain fiber—risk of CHD
6. sodium—hypertension
7. fruits and vegetables—cancer
8. sugar alcohols—dental caries
9. folic acid—neural tube defects (21 CFR Part 101 [1994])
10. soluble fiber—CHD

The Federal Trade Commission regulates advertising and requires that all advertising be true and not misleading.

Health Claims

No expressed or implied health claim may be made in the labeling for a food unless the claim is specifically authorized in the regulations, and the claim conforms to the requirements in such regulations (21 CFR § 101.14[e] [1996]) or is based on authoritative statements of appropriate government bodies (see under Labeling above). Infant formula and medical foods are exempt because they are covered under other provisions.

Food Additives

FFDCA is the primary statute governing food additives. FDA has issued numerous regulations and guidelines for manufacturers and users of food additives. Current information can be obtained from the FDA web page (http://www.fda.com).

Food additives are classified into two categories: (1) direct and (2) indirect. A food additive is defined by FDA, as "any substance, the intended use of which may reasonably be expected to result, directly or indirectly, in its becoming a component or otherwise affecting the characteristics of any food" (FFDCA § 201[s]: 21 CFR § 170.3[e]). Because food additives are not limited to ingredients manufacturers intend to use, food additives can also include lubricant oil used in food-processing equipment, lead paint used to decorate dishes, and polymers in food-packaging materials. It is the use to which the article is put, not the manufacturer's intent, that determines that a substance is a food additive. There are certain ingredients that are specifically excluded from the definition of food additive. Excluded ingredients are pesticides in or on raw agricultural commodities, color additives, prior-sanctioned substances, and generally regarded as safe (GRAS) substances.

The FDAMA of 1997 provided a number of changes to the procedures used by FDA. For example, FDAMA established premarket notification (PMN) procedures as the primary method by which the FDA regulates food additives that are food contact substances. FDAMA also permits health claims based on authoritative statements of appropriate government bodies in addition to those approved by specific regulation. Food contact substances include all substances that are intended for use as components of materials used in manufacturing, packing, packaging, transporting, or holding food if such use is not intended to have any technical effect on the food.

Food additives must meet two safety standards: (1) a general safety clause for noncarcinogens and the (2) Delaney Clause for carcinogens. Under the general safety clause, the law states that no regulation shall be issued if the data fail to establish that the proposed use of the additive, under the condition of use specified in the regulation, will be safe. Safety is further defined through FDA regulations as meaning a "reasonable certainty that the substance is not harmful under the intended conditions of use." The Delaney Clause, passed in the 1950s in response to a general concern by consumers regarding cancer-causing substances, states that "no additive shall be deemed to be safe if it is found to induce cancer when ingested by man or animal, or if it is found, after tests which are appropriate for the evaluation of the safety of food additives, to induce cancer in man or animal (FFDCA § 409[C][3][A]: 21 CFR §721 [b][5][B]).The Delaney Clause has been subject to administrative and judicial definitions over the past 40 years. In general, it has been interpreted that there is some level of cancer-causing substance which is *de minimus* and that FDA should not waste its regulatory resources on substances presenting *de minimus* risks. Typically, FDA has assumed that *de minimus* risks are those representing an incremental risk of cancer of less than one case in a million lives.

Color Additives

Color additives are regulated under provisions that are similar to those for food additives. However, there is no category that is generally regarded as safe for colors nor are there prior-sanctioned colors. There is a formal listing process for colors that includes the definition of the substances, defines the intended uses, and provides the specifications for the substances.

THE FOOD ADDITIVE APPROVAL PROCESS

The Office of Premarket Approval within the FDA reviews and approves food additive petitions. Foods and food additives are regulated differently than dietary supplements. Food additives are regulated under FFDCA, while supplements are regulated under legislation passed in 1994, the Dietary Supplement Health and Education Act (DSHEA, 1994).

Direct Food Additives

Direct food and color additives require certain steps for approval. Each specific use or category of use:

- Must be approved.
- Must demonstrate that it has an intended physical or technical effect.
- Must demonstrate safety under the proposed uses following the FDA publication, *Toxicological Principles for the Safety Assessment of Direct Food Additives and Color Additives Used in Food* (1982) (commonly referred to in the United States as *The Redbook*).

The applicant submits a petition to the FDA for its evaluation. The regulations are described in detail in 21 CFR § 170.20 (general principles for evaluating safety), § 170.22 (safety factors to be considered), § 170.38 (determination of food additive status), Part 171 (direct food additives), and Part 174 (indirect food additives). The petition must include proposed uses for the ingredient (by food or food category), estimates of dietary intake in relation to the acceptable dietary intake (ADI), *The Redbook* level of concern, complete toxicology studies, and an environmental impact assessment. FDA reviews the petition and presents its conclusions in the *Federal Register*. Although there are specified time limits for this review process, there are many delays, and approvals often take years. FDA has issued guidelines for the preparation of petition submissions, which are updated periodically. The most recent version can be obtained through the FDA web site (http://www.fda.gov).

Indirect Additives

Indirect food additives are either reviewed through a PMN procedure or through the petition procedure. The PMN procedure is new and currently under draft guidelines. In general, the information provided is similar for either the PMN or petition approach, but the review process is different. The PMN process can be used unless the FDA believes that the premarket review and approval are necessary to ensure adequate safety. FDA will not disclose the existence of a PMN for 120 days after submission. At the completion of the 120-day review period, the existence of the PMN and the information in a PMN will be publicly available with the exception of trade secret and confidential commercial information.

The PMN procedure is acceptable except under certain circumstances, such as the following:

- Uses of a food contact substance will increase the cumulative dietary concentration of the substance from food uses to greater than 1 part per million (ppm) or to greater than 200 parts per billion (ppb) in the case of biocides.
- The ADI for the food contact substance does not exceed the cumulative estimated daily intake by a factor of at least five.
- There are carcinogenicity studies that appear to have significant findings and that have not been reviewed by FDA.

Indirect additives are found in food as a result of their migration from packaging, and there is no technical effect in or on the food to which the substance migrates. There is a procedure, the Threshold of Regulation, that can be invoked for many indirect additives that shortens the approval time. Once the PMN process is finalized, it will replace the threshhold of regulation policy. Guidance for submitting requests under this procedure is provided for in 21 CFR § 170.39.

The Threshold of Regulation for substances used in food contact articles can be used only if the indirect additive is not a carcinogen and if migration in dietary concentrations is less than 0.5 ppb or less than 1% of the ADI. Further, there can be no technical effect in or on food to which the substance migrates, and there can be no adverse impact on the environment. Guidance regarding demonstrating that there is no adverse impact on the environment is available in 21 CFR Part 25, Environmental Impact Considerations.

A request for consideration under the Threshold of Regulation should include the following information:

- chemical composition of the substance for which the request is made, including the name of the chemical and CAS number
- intended technical effect of the substance in the food contact article (e.g., stabilizer or catalyst)
- detailed information on the conditions of use of the substance
- clear statement for the justification under the Threshold of Regulation (e.g., demonstration that the use results in dietary concentrations at or below 0.5 ppb or demonstration that the use involves a regulated direct food additive for which the dietary exposure is at or below 1 percent of the ADI)
- data to enable FDA to estimate the daily dietary concentration, such as validated migration data under worst-case (time/temperature) intended use conditions using appropriate food-simulating solvents (see *Recommendations for Chemistry Data for Indirect Food Additive Petitions* [FDA, 1995]), the level of the substance used in the manufacture of the food contact article, or the residual level of the substance present in the finished food contact article
- detailed description of the analytical method used to quantify the substance along with data used to validate the method
- for repeat-use articles, an estimate of the amount of food that contacts a specific unit of surface area over the lifetime of the article
- results of a literature search of existing toxicological information on the substance and its impurities, including whether long-term assays to assess carcinogenicity have been conducted and, if so, the results of those studies

- information that FDA can use to evaluate potential environmental impacts

Guidance is available in the 21 CFR § 25.31 (a)(b)(1) and (2). Additional information is contained in *Recommendations for Chemistry Data for Indirect Food Additive Petitions* (FDA, 1995). This can be obtained through the FDA web page listed above.

GRAS Substances

In 1997 FDA proposed a new system that would allow manufacturers to obtain much more rapid affirmation from FDA that a food substance is GRAS. Substances that are GRAS by qualified experts are not required by law to receive FDA approval before marketing. For substances that qualify under the GRAS regulations, manufacturers have long convened panels of qualified experts and essentially conducted self-affirmation of the GRAS status of the substance. Although this procedure met the legal requirements, it did not provide manufacturers with a positive affirmation from FDA. The new proposal is a simplified notification procedure that would replace the current lengthy rule-making process that would conserve resources by both FDA and manufacturers and should assure the consumer that the U.S. food supply is safe.

Products Designed To Decrease the Risk of Food-Borne Illness

FDA is currently expediting review for food additive products designed to decrease the risk of food-borne illness. Specifically, expedited review will be given to food additives intended to significantly decrease pathogenic strains of *E. coli, Salmonella, Campylobacter, Cyclospora,* and *Listeria* or their toxins. The petition must still meet the same approval standards applied to other food additive petitions, including valid scientific documentation of safety and efficacy (*Federal Register,* 1999).

REGULATION OF CONTAMINANTS AND PESTICIDES

Contaminants

Section 402 of the FFDCA prohibits adulteration of food. Unavoidable contaminants are addressed in the Good Manufacturing Practices regulations. FDA establishes defect action levels for poisonous or deleterious substances in human food and animal feed under section 402 of the FFDCA (21 CFR § 110). Action levels and tolerances are established based on the unavoidability of the poisonous or deleterious substances and do not represent permissible levels of contamination where it is avoidable. Action levels are established for some pesticides that are no longer registered to cover unavoidable residues due to persistence in the environment.

The blending of a food or feed containing a substance in excess of an action level or tolerance with another food or feed is not permitted, and the final product resulting from blending is unlawful, regardless of the level of the contaminant.

Action levels generally represent limits at or above which FDA will take legal action to remove products from the market. Where no established action level or tolerance exists, FDA may take legal action against the product at the minimal detectable level of the contaminant. The action levels are established and revised according to criteria specified in 21 CFR Parts 109 and 509 and are revoked when a regulation establishing a tolerance for the same substance and use becomes effective. Notices are published in the *Federal Register* as new action levels are established or as existing action levels are revised or revoked. It is the responsibility of the user of the list to keep up to date on changes in the action levels. FDA publishes a booklet periodically that can be obtained from the following address:

Industry Activities Staff (HFS-565)
CFSAN/FDA
200 C Street, S.W.
Washington, DC 20204
Telephone: (202) 205-5251

Copies of the Compliance Policy Guides referenced in the action level list may be purchased from:

National Technical Information Service
5285 Port Royal Road
Springfield, VA 22161

Pesticides

Overview of Current Regulations

The regulation of food and feed that contain pesticide residues is governed by sections 402, 408, and 409 of FFDCA. The Federal Insecticide, Fungicide and Rodenticide Act (FIFRA) requires EPA registration for all pesticides sold in the United States (FIFRA, 1996). FIFRA was revised and strengthened substantially in August 1996 by the FQPA (FQPA, 1996). These amendments fundamentally changed the way EPA regulates pesticides. The requirements of the new law include a new safety standard—reasonable certainty of no harm—that must be applied to all pesticides used on foods. The new law establishes a strong, health-based safety standard for pesticide residues in all foods that eliminates long-standing problems posed by multiple standards for pesticides in raw and processed foods.

FQPA requires EPA to consider all nonoccupational sources of exposure, including drinking water, and exposure to other pesticides with a common mechanism of toxicity when setting tolerances. FQPA contains special provisions for ensuring the safety of infants and children. The new law incorporates language virtually identical to the Clinton Administration's 1994 bill to implement key recommendations of the National Academy of Sciences (NAS) report, *Pesticides in the Diets of Infants and Children* (NAS, 1986). Briefly, FQPA requires an explicit determination that tolerances are safe for children; it also includes an additional safety factor of up to tenfold, if necessary, to account for uncertainty in data relative to children. It also requires consideration of children's special sensitivity and exposure to pesticides.

Unlike previous provisions of FIFRA that contained an open-ended provision for the consideration of pesticide benefits when setting tolerances, the new law places specific limits on benefit considerations. Benefit considerations apply only to nonthreshold effects of pesticides (e.g., carcinogenic effects); benefits cannot be taken into account for reproductive or other threshold effects. Benefits are further limited in three ways:

1. There is a limit on the acceptable risk in any one year.
2. There is a limitation on lifetime risk, which allows EPA to remove tolerances after specific phase-out periods.
3. Benefits cannot be used to override the health-based standards for children.

FQPA requires that all existing tolerances be reviewed within 10 years to make sure they meet the requirements of the new health-based safety standard. It also incorporates provisions for endocrine testing and provides new authority to require that chemical manufacturers provide data on their products, including data on potential endocrine effects.

FQPA includes enhanced enforcement of pesticide residue standards by allowing FDA to impose civil penalties for tolerance violations and requires distribution of a brochure in grocery stores on the health effects of pesticides, how to avoid risks, and which foods have tolerances for pesticide residues based on benefits considerations. The act also specifically recognizes a state's right to require warnings or labeling of food that has been treated with pesticides, such as California's Proposition 65. A state may not set tolerance levels that differ from national levels, however, unless the state petitions EPA for an exception, based on state-specific situations. National uniformity, however, would not apply to tolerances that included benefit considerations. FQPA establishes new requirements to expedite the review and registration of antimicrobial pesticides and ends regulatory overlap in jurisdiction over liquid chemical sterilants.

The full text of the FQPA is available from the EPA web site (www.epa.gov). Additional information on the implementation of FQPA can also be obtained from the EPA web site.

Established U.S. Tolerances for Pesticides

Section 408 of the FFDCA authorizes EPA to establish a tolerance for the maximum amount of a pesticide residue that may be legally present in or on a raw agricultural commodity. This section also authorizes EPA to exempt a pesticide residue in a raw agricultural commodity from the requirement of a tolerance. A tolerance or tolerance exemption is required when EPA grants registration under FIFRA for the use of a pesticide in food and feed production in the United States. Registration of a pesticide is not, however, a prerequisite for establishing a tolerance. For example, EPA may establish a temporary tolerance under section 408(j) to permit the experimental use of a nonregistered pesticide, or EPA may establish a tolerance for a pesticide residue resulting from the use of the pesticide in food or feed production in a foreign country. Tolerances and exemptions from tolerances established by EPA for pesticide residues in raw agricultural commodities are listed in 40 CFR Part 180.

Food Additive Regulations for Pesticides

A tolerance or tolerance exemption for a pesticide residue in a raw agricultural commodity also applies to the processed form of the commodity when ready to eat (§402[a][2][C] of the FFDCA). However, if a pesticide is to be used on a processed food or feed, or if residue present in or on a raw agricultural commodity in conformity with its tolerance under section 408 concentrates during processing to a level when ready to eat that is greater than the tolerance for the raw agricultural commodity, then a food additive regulation is required. In either instance, EPA is authorized under section 409 of the FFDCA to establish a food additive regulation for the maximum amount of a pesticide residue that may be legally present in a processed food or feed. Food additive regulations issued by EPA for pesticide residues in processed food and feed appear in 21 CFR Part 193 and in 21 CFR Part 561, respectively.

FDA is responsible for the enforcement of pesticide tolerances and food additive regulations established by EPA. This enforcement authority is derived from section 402(a)(2)(B) of the FFDCA. Under this section, a raw agricultural commodity or a processed food or feed is deemed to be adulterated and subject to FDA enforcement action when it contains either a pesticide residue at a level greater than that specified by a tolerance or food additive regulation or contains a pesticide residue for which there is no tolerance, tolerance exemption, or food additive regulation.

There are exceptions to FDA's enforcing an adulteration charge under section 402 for a pesticide residue in a food or feed that is not subject to a tolerance, tolerance exemption, or food additive regulation.

Unavoidable Pesticide Residues. Food or feed may contain a pesticide residue from sources of contamination that cannot be avoided by good agricultural or manufacturing practices, such as contamination by a pesticide that persists in the environment. In the absence of a tolerance, tolerance exemption, or food additive regulation, FDA may establish an action level for such unavoidable pesticide residues. An action level specifies the level below which FDA exercises its discretion not to take enforcement action. An action level established by FDA is based on EPA's recommendation, which follows the criteria of section 406 of the FFDCA. (See 21 CFR Parts 109 and 509 for information on FDA policy and procedures for establishing action levels for unavoidable food and feed contaminants.) Food or feed found to contain an unavoidable pesticide residue at a level that is at or greater than an action level is subject to FDA enforcement action.

EPA Emergency Exemptions. EPA is authorized by section 18 of FIFRA to grant an exemption from the registration requirements for the use of a nonregistered pesticide under emergency conditions. FQPA specified changes in the process for approval of emergency exception. EPA is to propose a level for a temporary tolerance and FDA is to be responsible for enforcement.

FDA will also consider taking enforcement actions for violation of sections 402(a)(2)(B) or 402(a)(2)(C) in the following situations:

- A food or feed contains residues of two or more pesticides of the same chemical class and the total amount of such residues when added together exceeds the lowest numerical tolerance for residues of one of the pesticides found in that class as set forth in 40 CFR § 180.3(e)(1). (In applying the criteria in this regulation, the residues to be added together must be at or above the analytical limit of quantitation as specified in the *FDA Pesticide Analytical Manual*, Volume I, §143.21 [FDA, 1998b]).
- A processed food or feed was derived from a raw agricultural commodity that contained a pesticide residue that did not conform to an established tolerance or tolerance exemption.
- In the absence of a food additive regulation and in accordance with 21 CFR § 170.19 or 21 CFR § 570.19, a pesticide residue in a processed food or feed when ready to eat is greater than the tolerance prescribed for the raw agricultural commodity.

Imported Foods and Feeds

The requirements of the FFDCA apply equally to domestically produced and imported food and feed. For example, even though the use of a pesticide in a foreign country is not subject to EPA registration requirements under FIFRA, a pesticide residue in imported food or feed must be in conformity with a tolerance, tolerance exemption, or food additive regulation established by EPA or, if the pesticide residue is unavoidable, an action level established by FDA. Although FDA formally establishes tolerances for pesticides, EPA is responsible for conducting the safety evaluations and recommending appropriate levels to FDA. Once tolerances have been established, FDA is responsible for monitoring the food supply and taking enforcement actions where appropriate.

FDA Action Levels for Unavoidable Pesticide Residues in Food and Feed Commodities

Action levels have been established for the following pesticides:

- Aldrin and Dieldrin
- benzene hexachloride (BHC)
- chlordane
- chlordecone (Kepone)
- DDT, DDE, and TDE
- dicofol (Kelthane)
- ethylene dibromide (EDB)
- heptachlor and heptacholor epoxide
- lindane
- Mirex

None of these action levels is binding on the agency, the regulated industry, or the courts. In any given case, FDA may decide to initiate an enforcement action below the action level or decide not to initiate an enforcement action if the level is exceeded. Unless otherwise specified, an action level listed for: (1) a raw agricultural commodity (other than grains) may also apply to the corresponding processed food intended for human consumption; (2) grains may also apply to both raw and processed grains intended for human or animal consumption; (3) fish may also apply to shellfish and processed fish intended for human consumption; and (4) processed animal feed may include mixed feeds and feed ingredients.

REFERENCES

Dietary Supplement Health and Education Act. (1994). Pub. L. No. 103-417. 108 Stat. 4325 (1994). Washington, DC: U.S. Government Printing Office.

Egg Products Inspection Act. (1970). Pub. L. No. 91-597. 84 Stat. 1620 (1970), 21 U.S.C. § 3031 et seq. Washington, DC: U.S. Government Printing Office.

Federal Food, Drug and Cosmetic Act. (1938). Pub. L. No. 75-717, 52 Stat. 2040 (1938) as amended 21 U.S.C. 301 et seq. (1994). Washington, DC: U.S. Government Printing Office.

Federal Insecticide, Fungicide and Rodenticide Act. (1996). As amended. Public Law No. 92-516, 92nd Congress, H.R. 10729, October 21, 1972, and August 6, 1996. Washington, DC: U.S. Government Printing Office.

Federal Meat Inspection Act. (1967). Pub. L. No. 59-242. 34 Stat. 12607. 21 U.S.C. § 601 et seq.

Federal Register. (1999). Vol. 64, 517.

Food and Drug Administration. (1982). *Toxicological Principles for the Safety Assessment of Direct Food Additives and Color Additives Used in Food* (also known as *The Redbook*). www.fda.gov. Washington, DC: Food and Drug Administration.

Food and Drug Administration. (1993). *Recommendations for Submission of Chemical and Technological Data for Direct Food Additive and GRAS Food Ingredient Petitions*. Washington, DC: Food and Drug Administration.

Food and Drug Administration. (1995). *Recommendations for Chemistry Data for Indirect Food Additive Petitions*. Washington, DC: Food and Drug Administration.

Food and Drug Administration. (1996). *Guidance for Submitting Petitions for Approval of Food Additives, Color Additives or Generally Recognized as Safe (GRAS) Substances*. Washington, DC: Food and Drug Administration.

Food and Drug Administration. (1998a). *Compliance Policy Guide*. http://www.fda.gov. Accessed December 1999.

Food and Drug Administration. (1998b). In *Pesticide Analytical Manual*. Vol. I, § 143.21. Washington, DC: Food and Drug Administration.

Food and Drug Law Institute. (1996). *Basic Outlines on Food Law and Regulations. A Collective Work by Top Legal and Regulatory Experts in the Food and Drug Field*. Washington, DC: The Food and Drug Law Institute.

Food Quality Protection Act. (1996). Pub. L. No. 104-170, 110 Stat. 1489. Washington, DC: U.S. Government Printing Office.

Food Safety and Inspection Service. (1997). Generic HACCP models and guidance materials available for review and comment. *Federal Register* 62, 32053–32054.

McKenna & Cuneo. (1996). *Summary and Analysis of the Food Quality Protection Act of 1996*. Washington, DC: McKenna & Cuneo, LLP, Publishers.

National Academy of Sciences. (1986). *Pesticides in the Diets of Infants and Children.* Washington, DC: National Academy of Sciences.

Nutrition Labeling and Education Act. (1990). Pub. L. No. 101-535. 104 Stat. (1990). Washington, DC: U.S. Government Printing Office.

Poultry Products Inspection Act. (1957). 21 U.S.C. § 451. Washington, DC: U.S. Government Printing Office.

U.S. Code of Federal Regulations 21 CFR §1–180. (These volumes describe in details the regulation of foods and food additives.)

CHAPTER 4

European Community Legislation on Limits for Additives and Contaminants in Food

Stuart A. Slorach

INTRODUCTION

This chapter summarizes the preparation and adoption of European Community (EC) legislation on limits for food additives, residues of pesticides and veterinary drugs, and other contaminants, such as heavy metals, mycotoxins, and nitrate.

EC procedures for preparing and adopting limits for additives and contaminants in foods are complex and may be difficult to understand for those not intimately involved in the work. The procedures involve, among others, the European Commission and its subsidiary bodies (e.g., working groups, standing committees, and scientific committees); the European Council of Ministers; the European Parliament; the 15 member states of the European Union (EU); and the food, drug, and chemical industries. Furthermore, different procedures are used for different classes of substance (e.g., the procedure for additives is different from that for pesticide residues), and the procedures have changed over the years and are still changing. For these reasons, only a brief outline of the procedures used for setting limits is given here and readers wishing to immerse themselves in the detail, of which there is plenty, should consult the references given at the end of the chapter.

All EC legislation and proposals for legislation are published in the *Official Journal of the European Communities*. Information on current EC legislation is available on the Internet at http://europa.eu.int/eur-lex. Legislative instruments are mainly of two types—regulations and directives. Regulations apply directly (verbatim) in all member states, whereas directives are transposed into the national legislation of the member states, and their implementation may therefore differ slightly from country to country.

The provisions of the Agreement on the Application of Sanitary and Phytosanitary Measures (SPS Agreement) and the Agreement on Technical Barriers to Trade (TBT Agreement) require the member states, who have signed these agreements, to notify any proposals for legislation that may result in barriers to international trade, including legislation on limits for additives and contaminants in food.

SCIENTIFIC ADVISORY COMMITTEES

The European Commission has long had a number of scientific advisory committees that provide it with expert advice in a variety of areas, including six committees related to food safety. The experts on these committees are expected to provide independent advice and not represent their countries or the organizations that employ them. The results of the deliberations of the scientific advisory committees are published and are now also available on the Internet at http://europa.eu.int/comm/dg24/health/sc/scf/report.en.html.

In 1997 the European Commission transferred responsibility for food safety to Directorate-General XXIV (Consumer Policy and Public Health—now referred to as SANCO) and restructured the system. This change followed criticism by the European Parliament of the Commission's handling of the bovine spongiform encephalopathy (BSE) crisis. The Commission established a Scientific Steering Committee, laid down mandates for eight scientific advisory committees (including scientific committees on *inter alia* food, veterinary public health, medicinal products and medical devices, and plants) (EC, 1997a), and appointed members to these committees by a new procedure. Appropriately qualified scientists were invited to declare their interest in becoming members of these expert committees, and the principles for selection of members of the committees were published. The aim of reforming the system was to increase the independence of the expert committees from national government and other interests, and to increase transparency of the committee's work. Members of the expert committees are required to declare any interests they may have that could affect their impartiality in dealing with a particular subject or substance on a committee's agenda.

The committees most involved in setting limits for additives and contaminants in food are the Scientific Committee for Food (SCF), the Scientific Committee on Veterinary Measures Relating to Public Health, and the Scientific Committee on Plants. Prior to autumn 1997, there was a separate Scientific Committee for Pesticides, but its responsibilities are now included in the mandate of the new Scientific Committee on Plants. In order to facilitate their work, many of the scientific committees have subsidiary working groups that carry out a lot of the preparatory work that needs to be done (e.g., preparation of working documents) before decisions can be made by the committee. For example, SCF has working groups on additives, contaminants, flavorings, food contact materials, nutrition and diabetic foods, novel foods and processes, intake and exposure, and food hygiene and microbiology.

FOOD ADDITIVES

Council directive 89/107 (EC, 1989) provided the framework for EC legislation on food additives and general criteria for their use. However, this directive does not apply to processing aids, flavorings, or substances added as nutrients. The main goal of EC legislation in this area was to remove barriers to trade within the EC caused by differences in national legislation on additives and conditions for their uses. Detailed controls on the use of food additives, including maximum levels of use, are to be found in three European Parliament and European Council directives on sweeteners (directive 94/35 [EC, 1994a]), colors (directive 94/36 [EC, 1994b]), and additives other than colors and sweeteners (directive 95/2 [EC, 1995a]). Some of these directives have subsequently been amended, and further amendments will no doubt be made. All amendments are published in the *Official Journal of the European Communities*.

Directives laying down criteria for the identity and purity of certain groups of food additives (e.g., sweeteners, colors, and emulsifiers) have been issued. Work on establishing such criteria for the remaining groups of additives is expected to be completed in the next few years. It has been suggested that additives produced by using genetically modified organisms be reevaluated by the SCF.

The general criteria for the use of food additives require that the additives present no hazard to the health of the consumer at the proposed level of use and that the amount of the additive used should be limited to the lowest level necessary to achieve the desired effect. Approval of a food additive must also take into account any acceptable daily intake (ADI) for man, or equivalent assessment, established for the additive, and the probable daily intake of the additive from all sources. When the additive is to be used in foods eaten by special groups of consumers, account should be taken of the probable daily intake of the additive by consumers in such groups.

Requests for approval of new food additives or changes in the conditions of use of additives that are already approved are submitted to the European Commission by the member states. The Commission considers the request and prepares a proposal for new legislation, which is then reviewed by *inter alia* several committees (e.g., the SCF and the Economic and Social Committee) as well as the European Council and the Parliament in a lengthy and complex procedure. Following review, the initial proposal may be amended and subject to further comment.

If proposed changes to the food additive regulations could have public health implications, the European Commission is required to ask the SCF for its opinion. These opinions are published. The SCF recommends ADIs for food additives. These ADIs are usually similar or identical to those proposed by the Joint Food and Agriculture Organization/World Health Organization (FAO/WHO) Expert Committee on Food Additives (JECFA), which provide the toxicological basis for Codex Alimentarius Commission (CAC) standards on food additives.

In the detailed EC legislation on colors, sweeteners, and other food additives (see the above-mentioned directives), the use of some additives is restricted to a certain food or group of foods, whereas other additives may be used much more widely. The maximum permitted levels for some additives of low toxicity (e.g., sorbitol) are expressed as *quantum satis* (i.e., no maximum level is specified). In such cases, however, the additives must be used in accordance with good manufacturing practice at a level not higher than that necessary to achieve the intended purpose, and their use must not result in deception of the consumer. For other additives, for example most colors, antioxidants, and preservatives, the maximum level of the additive permitted in each food or group of foods is specified in mg/kg or mg/l in the appendixes to the directives. The starting point for establishing maximum permitted levels is usually information on technological need supplied by the additive manufacturers and/or the food industry. When reviewing proposals for legislation, some member states try to estimate whether the proposed maximum levels are compatible with the desire to keep the intake of the food additive from all sources below the ADI recommended by the SCF. This is particularly important for additives with low ADIs and additives for which the current intake is estimated to be close to the ADI.

The detailed directives on food additives require the member states to establish a system of consumer surveys to monitor additive consumption and require the European Commission to submit reports on the information obtained to the European Parliament and the European Council within a specified time. On the basis of these reports, it is decided whether it is necessary to introduce further restrictions on the use of additives in order to keep their intake below the ADIs recommended by the SCF.

Interaction with Codex

The EC member states and the European Commission take an active part in Codex work on food additives, especially in the Codex Committee on Food Additives and Contaminants, which is hosted by the Netherlands. The United Kingdom was instrumental in promoting the adoption of the so-called *horizontal approach* to food additive standards within Codex that led to the development of the Codex General Standard for Food Additives. In addition, the proposed CAC list of foods to which no additives should be added is essentially the same as that adopted earlier by the EC (see Chapter 11, The Codex General Standard for Food Additives—A Work in Progress).

PESTICIDE RESIDUES

The procedures used to set permanent maximum residue levels (MRLs) for pesticides are described briefly below. In parallel with this, provisional MRLs are prepared in connection with the work on regulating the placing of plant protection products on the market according to European Council directive 91/414 (EC, 1991a).

Toxicological Evaluation

A subgroup of the European Commission's Working Group on Plant Health proposes MRLs for pesticide residues in foods. For each pesticide, MRLs are proposed for residues in the relevant individual commodities/crops. Since a large number of pesticides have to be dealt with, the workload is spread by appointing each member state in the working group as rapporteur for a number of specified pesticides.

When proposing an MRL, the rapporteur member state (RMS) identifies the ADI that is valid for the pesticide in question. The ADI identified is often the same as that recommended by the Joint FAO/WHO Meeting on Pesticide Residues (JMPR), whose recommendations on ADIs and MRLs are used within the CAC system (see Chapter

13, Codex Standards for Pesticide Residues). If the ADI proposed by the RMS is not the ADI recommended by JMPR, the RMS has to provide an explanation for the difference. The other member states comment on the RMS proposal at meetings of the Working Group on Pesticide Residues. If the member states cannot reach an agreement on the evaluation, the matter is referred to one or more of the European Commission's scientific committees. Prior to autumn 1997, such questions were referred to the Scientific Committee for Pesticides, but since then they are referred to the SCF or the Scientific Committee on Plants, or both.

Residue Data

Data on the levels of residues found in supervised trials, in which the pesticide is used according to good agricultural practice (GAP), are the main basis for proposing MRLs. Such trials must include studies that reflect the use that leads to the highest residue levels or critical GAP. MRLs are established for individual crops or groups of crops. In certain cases, data from supervised trials on one crop may be extrapolated to another crop. The results of the supervised trials are evaluated by the RMS and then discussed by the Working Group on Pesticide Residues. Previously there was some reluctance to use residue data generated in studies carried out outside Europe. However, in recent years such data have been considered.

Dietary Exposure Calculations

Although the main basis for proposing MRLs is the data from supervised trials, when considering the proposals, member states calculate the theoretical maximum daily intake that could occur if the proposed MRLs were adopted, using the WHO European Diet, or national dietary information, as the basis for their calculations. If the theoretical maximum daily intake exceeds the proposed ADI, member states may not be prepared to accept the proposed MRL without further refinement of the intake calculations. The EC considers not only chronic exposure to pesticide residues via foods, but also acute exposure. The calculation of acute exposure is presently based on the recommendations from the Joint FAO/WHO Expert Consultation on Food Consumption and Intake Assessment of Chemicals, held in Geneva, February 10–14, 1997.

Preparation and Adoption of EC Legislation

EC MRLs are laid down according to procedures in directive 97/41 (EC, 1997c), which amended directives 76/895 (EC, 1976), 86/362 (EC, 1986a), 86/363 (EC, 1986b), and 90/642 (EC, 1990a). Earlier decisions on MRLs were made by a procedure in which the European Council made the final decision. Now decisions are usually made by the European Commission according to a regulatory committee decision procedure (Procedure IIIb) involving the Commission's Standing Committee on Plant Health. This standing committee consists of representatives of the member states but is chaired by the Commission. If the members of this standing committee cannot agree (i.e., there is not a qualified majority for a proposal), the question is referred to the Council for a decision. The Council must act by a qualified majority. (A qualified majority results from the use of a weighted voting system where member states each have a number of votes; a simple majority arises when member states each have one vote.) If the Council fails to reply within a stipulated time limit, the proposal will be adopted by the Commission, except where the Council has decided against the measure by a simple majority. Before the proposals for MRLs are notified to the World Trade Organization, they are sent to the Scientific Committee on Plants or the SCF, or both.

Community Legislation on MRLs of Pesticides

European Council directive 76/895 (EC, 1976) (as amended by directives 81/36 [EC, 1981], 82/528 [EC, 1982], and 88/298 [EC, 1988]) contains recommendations for MRLs for pesticides in or on fruit and vegetables. However, these MRLs are not mandatory and member states may set higher, but not lower, MRLs in their national legislation.

European Council directives 86/362 (EC, 1986a), 86/363 (EC, 1986b), and 90/642 (EC, 1990a) contain MRLs for pesticides in or on cereals, foods of animal origin, and fruit and vegetables, respectively. These MRLs are mandatory and must be incorporated into the national legislation of the member states. These directives have been amended several times. Directives 88/298 (EC, 1988), 93/57 (EC, 1993a), 94/29 (EC, 1994c), 95/39 (EC, 1995b), 96/33 (EC, 1996a), 98/82 (EC, 1998a), and 99/71 (EC, 1999d) concern MRLs for cereals and foods of animal origin, while directives 93/58 (EC, 1993b), 94/30 (EC, 1994d), 95/38 (EC, 1995c), 95/61 (EC, 1995d), 96/32 (EC, 1996b), 98/82 (EC, 1998a), and 99/71 (EC, 1999d) concern MRLs for fruit and vegetables. Special EC limits for pesticide residues in foods for infants and young children have been laid down in European Commission directives 99/39 and 99/50 (EC, 1999b and 1999c). Unfortunately, official consolidated versions of the amended directives are not produced, which makes it difficult to get a complete picture of all the MRLs that have been adopted to date.

Interaction with the Codex Committee on Pesticide Residues

Many member states participate actively in the work of this committee that is hosted by the Netherlands. Before and during each meeting of the Codex Committee on Pesticide Residues (CCPR), EC positions are coordinated as much as possible. In view of the special status of CAC MRLs since the signing of the SPS Agreement, the EC now attaches much more importance to Codex work. In recent compilations of proposals for MRLs, the European Commission has also included the corresponding CAC MRLs for comparison.

VETERINARY DRUG RESIDUES

Within the EC, the MRL for a veterinary drug residue is defined as "the maximum concentration of residue resulting from the use of a veterinary medicinal product (expressed in milligrams or micrograms per kilogram on a fresh weight basis) which may be accepted by the Community to be legally permitted or recognized as acceptable in or on food (EC, 1990b)."

The procedures used within the EC to establish MRLs for veterinary medicinal products in foods of animal origin are laid down in European Council regulation 90/2377 (EC, 1990b), which came into force on January 1, 1992. The most important principle in these procedures is that foods obtained from treated animals must not contain residues that might constitute a health hazard for the consumer.

When calculating MRLs, the aim is to ensure that the total daily intake of residues of the substance via foods of animal origin, averaged over a lifetime, does not exceed the ADI. For the purposes of these calculations, the body weight of the consumer has been assumed to be 60 kg, and the consumption of various foods has been assigned certain values (e.g., milk,1500 g; muscle, 300 g; liver, 100 g). In cases where a substance is also used as a pesticide, the exposure from such use is also taken into account. The procedures described below apply to veterinary drugs used as medicines, but not to feed additives used for growth promotion purposes.

MRLs are determined by the Committee for Veterinary Medicinal Products (CVMP), and its Safety of Residues Working Party (SRWP), attached to the European Medicine Evaluation Agency (EMEA). Different procedures are used, depending on when the pharmacologically active substance was authorized for use. For medicines containing substances authorized after January 1, 1992 (i.e., new substances), MRLs must be set at the European level for all pharmacologically active substances before approval procedures can be started in the member states. MRLs for medicines authorized before January 1, 1992 (i.e., old substances) must be evaluated by December 31, 1999, if their use after that date is to continue. Regardless of the procedure to be followed in setting the MRL, the manufacturer of a veterinary medicinal product must provide safety and residue dossiers containing the information required to set an ADI and MRLs. The safety dossier contains the pharmacodynamic, kinetic, metabolic, and toxicity data, and the residue dossier contains data on kinetics, metabolism, and residues, as well as the analytical method(s) for the substance.

Regulation 90/2377 (EC, 1990b) contains the following four annexes in which the substances are listed after evaluation:

1. substances for which final MRLs have been established
2. substances for which MRLs are not deemed necessary in order to protect public health
3. substances with provisional MRLs (if a dossier is incomplete, the manufacturer may be given a set time (up to five years) in which to provide the necessary information)
4. substances for which it is not possible, for various reasons, to set an MRL (i.e., the administration of such substances is prohibited throughout the EU

and the marketing authorization for the medicines concerned has been withdrawn)

MRLs for Substances Authorized after January 1, 1992

Applications from industry are submitted to EMEA. Since a large number of substances have to be dealt with, the workload is spread out by appointing each member state represented in the CVMP as rapporteur or corapporteur for a number of specified veterinary drugs. Using information in the dossier provided by the manufacturer, the RMS proposes an ADI for the drug in question and MRLs for relevant foods of animal origin (e.g., muscle, liver, and milk). This is done using the guidelines in regulation 90/2377 (EC, 1990b) and in the *Rules Governing Medicinal Products in the European Community*, Volume VI (EC, 1991c). The ADI and MRLs are often, but not always, the same as those proposed by JECFA or JMPR for the same drug. Other member states in the CVMP then comment on the ADI and MRLs proposed by the RMS. When the CVMP has reached agreement on MRLs for a drug, they are submitted to the European Commission, for adoption by the Committee for the Adaptation to Technical Progress of the Directives on the Removal of Technical Barriers to Trade in the Veterinary Medical Products Sector (now called the Standing Committee for Veterinary Medicinal Products). If a qualified majority of the member states in that committee supports adoption, the MRLs are then incorporated into the relevant annex to regulation 90/2377 (EC, 1990b). If a qualified majority is not obtained, the Commission proposes to the European Council the measures to be adopted; the Council must act by a qualified majority. If the Council has not acted within three months, the proposed measures are adopted by the Commission, unless the Council has voted against them by a simple majority. All amendments to the annexes of regulation 90/2377 (EC, 1990b) are published in the *Official Journal of the European Communities*.

MRLs for Substances Authorized before January 1, 1992

After examining dossiers supplied by industry, RMPs in the CVMP's SRWP propose ADIs and MRLs for substances authorized before January 1, 1992. These proposals are then discussed by the CVMP. When the CVMP has reached agreement on MRLs, they are then adopted by the procedure described above for new substances.

Withdrawal Periods

The withdrawal period is the time between the last dose given to the animal and the time when the level of residues in the tissues (e.g., muscle, liver, kidney, and skin/fat) or products (e.g., milk, eggs, and honey) is lower than or equal to the MRL. For veterinary medicinal products intended to be marketed in only one member state, withdrawal periods are set at the national (member state) level. For products intended to be used in more than one member state, a mutual recognition procedure has been obligatory since January 1, 1998. In this procedure, evaluation is carried out in one country and the proposed withdrawal period is then accepted (or rejected) by other member states. If the proposal is not accepted, the matter goes to arbitration at EMEA. For products intended to be marketed throughout the whole of the EU, withdrawal periods are determined by the CVMP by a central procedure analogous to that used for MRLs. This procedure must always be used for certain special groups of substances (i.e., those produced by biotechnology and innovative products).

Interaction with the Codex Committee on Residues of Veterinary Drugs in Food

The member states and the European Commission take an active part in Codex work on MRLs for veterinary drugs, especially at the Codex Committee on Residues of Veterinary Drugs in Food (CCRVDF). Before and during each CCRVDF meeting, EC positions are coordinated as much as possible. As for residues of pesticides, the EC now assigns much more importance to CAC work in view of the special status attached to CAC MRLs since the signing of the SPS Agreement. The MRLs proposed by the JECFA, which reviews veterinary drug residues, and the JMPR are discussed in CCRVDF. In many, but not all, cases, these CAC MRLs are accepted by the member states of the EU. The most notable exceptions to this in recent years were the MRLs for hormones used for growth promotion, which have been the subject of much acrimonious debate in Codex and also the subject of an SPS dispute panel.

OTHER CONTAMINANTS

In addition to MRLs for residues of pesticides and veterinary drugs, the EC has, or is preparing, legislation concerning certain other contaminants, such as mycotoxins (e.g., aflatoxins and ochratoxin A), nitrate, and heavy metals (e.g., mercury, lead, and cadmium) in foods.

General Procedure

Council regulation 93/315 (EC, 1993c) lays down EC procedures for contaminants in food. The SCF must be consulted on all questions that may have an effect on public health. This committee of independent scientists carries out the toxicological evaluations that underpin the limits set for contaminants. The scientific data that form the basis of the evaluations are obtained mainly from the scientific literature and member states. Data on human exposure to contaminants, such as nitrates, cadmium, aflatoxins, and ochratoxin A, have been collected and collated in projects in a program on scientific cooperation between the member states (known as SCOOP).

Proposals for new limits prepared by European Commission working parties are submitted to the Standing Committee for Foodstuffs, which consists of representatives of the member states, but is chaired by the Commission. Decisions on new limits are usually made by the Commission according to a regulatory committee procedure (Procedure IIIb—for details see "Preparation and Adoption of EC Legislation" above). The Commission publishes the limits as a regulation in the *Official Journal of the European Communities*. Methods of sampling and analysis to check compliance with the maximum levels are also published.

Mercury and Histamine in Fishery Products

Council directive 91/493 (EC, 1991b), as amended by directive 95/71 (EC, 1995e), lays down the health conditions for the production and marketing of fishery products. Chapter V of that directive contains, among other things, maximum limits for histamine in certain fish species and instructions for checking that these limits are not exceeded. That chapter also makes provision for the establishment of limits for the presence in fish of contaminants from the aquatic environment.

Commission decision 93/351 (EC, 1993d) lays down maximum limits for mercury in fishery products and methods of sampling and analysis to check compliance with these limits. This decision was made after consulting the Standing Veterinary Committee. The mean total mercury content of the edible parts of fishery products must not exceed 0.5 mg/kg of fresh weight. However, this average limit is increased to 1 mg/kg fresh weight for the edible parts of certain species listed in the annex to the decision. The higher limit applies to *inter alia* sharks, tuna, swordfish, halibut, and pike.

Nitrate in Lettuce and Spinach

Proposals for limits on nitrates in certain vegetables were prepared by the European Commission's Committee of Experts in the Working Party on Agricultural Contaminants. The proposals were then considered using the above-mentioned procedure. The Commission issued regulation 194/97 (EC, 1997b), which sets maximum levels for nitrates in lettuce and spinach. This has later been amended by regulation 864/1999 (EC, 1999a).

Mycotoxins

Aflatoxins, ochratoxin A, and patulin have been evaluated toxicologically by the SCF. The question of maximum levels for these mycotoxins in foods has been discussed for several years in the Committee of Experts in the Working Party on Agricultural Contaminants under DG VI. Maximum levels for aflatoxin M1 in milk, and for aflatoxin B1 and the sum of aflatoxins B1, B2, G1, and G2 in ground nuts and certain other foods were laid down in European Commission regulation 98/1525 (EC, 1998b) and came into force on January 1, 1999. The question of maximum levels for ochratoxin A and patulin is still under discussion, and a new toxicological evaluation of the former substance is being carried out by the SCF.

Heavy Metals

Discussions on limits for lead and cadmium in foods have been going on for several years in a European Commission working party. The SCF has carried out toxicological evaluations on lead and cadmium. However, as yet no proposals for limits for these metals have been adopted by the Standing Committee for Foodstuffs.

Interaction with the Codex Committee on Food Additives and Contaminants

The contaminants reviewed above are mainly dealt with by the Codex Committee on Food Additives and Contaminants (CCFAC). Many of the member states of the EU are very active in Codex work on contaminants. For example, Denmark and the Netherlands have been instrumental in developing the Codex General Standard on Contaminants and Toxins, and draft limits for lead in various foods. Sweden has developed a proposal for a limit for ochratoxin A in cereals and cereal products.

CONCLUSIONS AND FUTURE DEVELOPMENTS

The EC has already established procedures for preparing and adopting legislation on limits for additives and contaminants in foods. In recent years, the work of the scientific advisory committees that carry out the health risk assessments and provide the scientific basis for most of the limits has become more independent and transparent; this trend is likely to continue. The procedures for adopting legislation are in many cases very complex and time-consuming, especially if the European Parliament is consulted. Whether more decision making will be delegated to the European Commission and the Council or the involvement of the Parliament will be extended remains to be seen.

Few radical changes are likely to occur in the next few years regarding limits for food additives. This is because few new additives are being produced and also because the current EC regulations on maximum levels of use are so liberal that, with few exceptions, the food industry does not require higher levels. However, the permitted areas of use of some additives may be extended to account for special products traditionally produced in new member states.

Much work still remains to be done on limits for pesticide residues. In addition to new substances, there is an urgent need to reevaluate many of the older pesticides in the light of new toxicological data. A timetable for the work planned for the next few years has been agreed. Continued cooperation with countries outside the EU should expedite matters.

Much work has still to be done on the preparation and adoption of maximum levels for mycotoxins, heavy metals, and other contaminants, such as dioxins and PCBs. Here one of the main factors delaying progress is the lack of data for toxicological evaluations and hence the setting of ADIs or tolerable daily or weekly intakes. Furthermore, there is a lack of good data on levels of contaminants in individual foods and on dietary intakes.

The member states already play an active role in the development of CAC limits for additives and contaminants. It is foreseen that this will continue and that coordination between the member states and the European Commission on Codex matters will further improve. Limits for many substances mentioned above (e.g., ochratoxin A, lead, cadmium, and some pesticides) are being discussed in parallel in the EC and in Codex and often by the same people. Such discussion facilitates coordination of the work in these different fora and should expedite the establishment of limits that can be widely accepted.

REFERENCES

European Community. (1976). Council directive 76/895/EEC of 23 November 1976 relating to the fixing of maximum levels for pesticide residues in or on fruit and vegetables. *Off J Eur Communities* L 340, 26–31.

European Community. (1981). Council directive 81/36/EEC of 9 February 1981 amending Annex II to directive 76/895/EEC relating to the fixing of maximum levels for pesticide residues in and on fruit and vegetables. *Off J Eur Communities* L 46, 33–34.

European Community. (1982). Council directive 82/528/EEC of 19 July 1982 amending Annex II to directive 76/895/EEC relating to the fixing of maximum levels for pesticide residues in and on fruit and vegetables. *Off J Eur Communities* L 234, 1–4.

European Community. (1986a). Council directive 86/362/EEC of 24 July 1986 on the fixing of maximum levels for pesticide residues in and on cereals. *Off J Eur Communities* L 221, 37–42.

European Community. (1986b). Council directive 86/363/EEC of 24 July 1986 on the fixing of maximum levels for pesticide residues in and on foods of animal origin. *Off J Eur Communities* L 221, 43–47.

European Community. (1988). Council directive 88/298/EEC of 16 May 1988 amending Annex II to directives 76/895/EEC and 86/362 relating to the fixing of maximum levels for pesticide residues in and on fruit and vegetables and cereals, respectively. *Off J Eur Communities* L 126, 53–54.

European Community. (1989). Council directive 89/107/EEC of 21 December 1988 on the approximation of the laws of the member states concerning food additives authorised for use in foodstuffs intended for human consumption. *Off J Eur Communities* L 40, 27–33.

European Community. (1990a). Council directive 90/642/EEC of 27 November 1990 on the fixing of maximal levels for pesticide residues in and on certain products of plant origin, including fruit and vegetables. *Off J Eur Communities* L 350, 71–79.

European Community. (1990b). Council regulation 90/2377/EEC of 26 June 1990 laying down a Community procedure for the establishment of maximum residue limits of veterinary medicinal products in foodstuffs of animal origin. *Off J Eur Communities* L 224, 7–14.

European Community. (1991a). Council directive 91/414/EEC of 15 July 1991 concerning the placing of plant protection products on the market. *Off J Eur Communities* L 230, 1–32.

European Community. (1991b). Council directive 91/493/EEC of 22 July 1991 laying down the health conditions for the production and placing on the market of fishery products. *Off J Eur Communities* L 268, 15–34.

European Community. (1991c). *The Rules Governing Medicinal Products in the European Community*, Vol VI. Establishment by the European Community of maximum residue limits (MRLs) for residues of veterinary medicinal products in foodstuffs of animal origin. Luxembourg: Office for Official Publications of the European Communities.

European Community. (1993a). Council directive 93/57/EEC of 29 June 1993 amending the annexes to directives 86/362/EEC and 86/363/EEC on the fixing of maximum levels for pesticide residues in and on cereals and foodstuffs of animal origin, respectively. *Off J Eur Communities* L 211, 1–5.

European Community. (1993b). Council directive 93/58/EEC of 29 June 1993 amending annex II to directive 76/895/EEC relating to the fixing of maximum levels for pesticide residues in or on fruit and vegetables and the annex to directive 90/462/EEC relating to the fixing of maximum levels for pesticide residues in and on certain products of plant origin, including fruit and vegetables, and providing for the establishment of a first list of maximum levels. *Off J Eur Communities* L 211, 6–39.

European Community. (1993c). Council regulation 93/315/EEC of 8 February 1993 laying down Community procedures for contaminants in food. *Off J Eur Communities* L 37, 1–3.

European Community. (1993d). Commission decision 93/351/EEC of 19 May 1993 determining analysis methods, sampling plans and maximum limits for mercury in fishery products. *Off J Eur Communities* L 144, 23–24.

European Community. (1994a). European Parliament and Council directive 94/35/EC of 30 June 1994 on sweeteners for use in foodstuffs. *Off J Eur Communities* L 237, 3–12.

European Community. (1994b). European Parliament and Council directive 94/36/EC of 30 June 1994 on colours for use in foodstuffs. *Off J Eur Communities* L 237, 13–29.

European Community. (1994c). Council directive 94/29/EC of 23 June 1994 amending the annexes to directives 86/362/EEC and 86/363/EEC on the fixing of maximum levels for pesticide residues in and on cereals and foodstuffs of animal origin, respectively. *Off J Eur Communities* L 189, 67–69.

European Community. (1994d). Council directive 94/30/EC of 23 June 1994 amending annex II to directive 90/642/EEC relating to the fixing of maximum levels for pesticide residues in and on certain products of plant origin, including fruit and vegetables, and providing for the establishment of a list of maximum levels. *Off J Eur Communities* L 189, 70–83.

European Community. (1995a). European Parliament and Council directive 95/2/EC of 20 February 1995 on food additives other than colours and sweeteners. *Off J Eur Communities* L 61, 1–40.

European Community. (1995b). Council directive 95/39/EC of 17 July 1995 amending the annexes to directives 86/362/EEC and 86/363/EEC on the fixing of maximum levels for pesticide residues in and on cereals and foods of animal origin. *Off J Eur Communities* L 197, 29–31.

European Community. (1995c). Council directive 95/38/EC of 17 July 1995 amending annexes I and II to directive 90/642/EEC on the fixing of maximum levels for pesticide residues in and on certain products of plant origin, including fruit and vegetables, and providing for the establishment of a list of maximum levels. *Off J Eur Communities* L 197, 14–28.

European Community. (1995d). Council directive 95/61/EC of 29 November 1995 amending annex II to directive 90/642/EEC on the fixing of maximum levels for pesticide residues in and on certain products of plant origin, including fruit and vegetables. *Off J Eur Communities* L 292, 27–30.

European Community. (1995e). Council directive 95/71/EC of 22 December 1995 amending the annex to directive 91/493/EEC laying down the health conditions for the production and placing on the market of fishery products. *Off J Eur Communities* L 332, 40–41.

European Community. (1996a). Council directive 96/33/EC of 21 May 1996 amending annexes to directives 86/362/EEC and 86/363/EEC on the fixing of maximum levels for pesticide residues in and on cereals and foodstuffs of animal origin. *Off J Eur Communities* L 144, 35–38.

European Community. (1996b). Council directive 96/32/EC of 21 May 1996 amending the annex II to directive 76/895/EEC relating to fixing of maximum limits for pesticide residues in and on fruit and vegetables and annex II to directive 90/642/EEC relating to the fixing of maximum levels for pesticide residues in and on certain products of plant origin, including fruit and vegetables, and providing for the establishment of a list of maximum levels. *Off J Eur Communities* L 144, 12–34.

European Community. (1997a). Commission decision 97/579/EC of 23 July 1997 setting up scientific committees in the field of consumer health and food safety. *Off J Eur Communities* L 237, 18–23.

European Community. (1997b). Commission Regulation 194/97/EC of 31 January 1997 setting maximum levels for certain contaminants in foodstuffs. *Off J Eur Communities* L 31, 48–50.

European Community. (1997c). Council directive 97/41/EC of 25 June 1997 amending directives 76/895/EEC, 86/362/EEC, 86/363 EEC and 90/462/EEC relating to the fixing of maximum levels for pesticide residues in and on, respectively, fruit and vegetables, cereals, foods of animal origin, and certain products of plant origin, including fruit and vegetables. *Off J Eur Communities* L 184, 33–49.

European Community. (1998a). Commission directive 98/82/EC of 27 October 1998 amending the annexes to Council directives 86/362/EEC, 86/363/EEC and 90/642/EEC on the fixing of maximum levels for pesticides in and on cereals, foodstuffs of animal origin and certain products of plant origin, including fruit and vegetables respectively. *Off J Eur Communities* L 290, 25–54.

European Community. (1998b). Commission regulation 98/1525/EC of 16 July 1998 amending regulation (EC) No. 194/97 of 31 January 1997 setting maximum levels for certain contaminants in foodstuffs. *Off J Eur Communities* L 201, 43–46.

European Community. (1999a). Commission regulation 864/1999 of 26 April 1999 amending regulation 194/97 setting maximum levels for certain contaminants in foodstuffs. *Off J Eur Communities* L 108, 16–18.

European Community. (1999b). Commission directive 99/39/EC of 6 May 1999 amending directive 96/5/EC on processed cereal-based foods and baby foods for infants and young children. *Off J Eur Communities* L 124, 8–10.

European Community. (1999c). Commission directive 99/50/EC of 25 May 1999 amending directive 91/321/EEC on infant formulae and follow-on formulae. *Off J Eur Communities* L 139, 29–31.

European Community. (1999d). Commission directive 99/71/EC of 14 July 1999 amending the annexes to council directives 86/362/EEC, 86/363/EEC, and 90/642/EEC on the fixing of maximum levels for pesticide residues in and on cereals, foodstuffs of animal origin, and certain products of plant origin, including fruit and vegetables respectively. *Off J Eur Communities* L 194, 36–44.

CHAPTER 5

Development of Australia/ New Zealand Standards

Simon Brooke-Taylor and Peter J. Abbott

INTRODUCTION

The Australia New Zealand Food Authority (ANZFA) is a statutory authority responsible for the development of food standards for Australia and New Zealand. The process by which ANZFA develops standards is prescribed in law and is intended to provide for open and transparent decision making. Standards developed by ANZFA are subject to approval by a council of ministers representing the Australian Commonwealth, State, and Territory Governments as well as the New Zealand Government.

ANZFA is currently engaged in an extensive review of the Australian Food Standards Code, scheduled to be completed by January 1, 2000. This review will result in the development of the Joint Australia New Zealand Food Standards Code, which will form the basis of a single market for food between the two countries.

Both Australia and New Zealand were active participants in the Uruguay Round of the General Agreement on Tariffs and Trade (GATT) negotiations, which led to the formation of the World Trade Organization (WTO). Consequently, both countries place a high level of importance on the principles embodied in the WTO trade agreements, especially those relating to the role of risk analysis in standards setting.

The development of standards by ANZFA is discussed in this chapter with particular regard to the risk analysis and the development of standards for food additives and contaminants.

FOOD REGULATION IN AUSTRALIA

Under Australia's federal system of government, food legislation is generally considered to be the legislative responsibility of the states and territories; each state and territory government is responsible for food sold within its boundaries. This situation has led to different safety criteria and labeling requirements being applied in different parts of the country.

In 1953 the National Health and Medical Research Council (NHMRC) assumed responsibility for the evaluation of food additives and the development of food standards at the national level. In 1980 in the interests of uniformity, the states and territories endorsed the development of a Model Food Act that could be implemented through state/territory legislation (NHMRC, 1983). Despite its name, the Model Food Act is not legislation in its own right, but sets out a blueprint for states and territories to introduce uniform food legislation within their own jurisdictions and administrative structures.

The Model Food Act is variously implemented in state/territory legislation (Exhibit 5–1). Of particular note in the context of food standards, the Model Food Act prescribes offenses in connection with the preparation, packaging, and sale of

- food that is unfit for human consumption
- food that is adulterated
- food that is damaged, deteriorated, or perished

The Model Food Act also addresses packaging, labeling, and advertising with respect to false and misleading claims; and the sale of food that is not of the nature, substance, or quality demanded by the purchaser. The Model Food Act as implemented also provides the means by which food standards are enforced under state and territory food legislation.

From 1982 NHMRC standards were consolidated in the NHMRC document *Approved Food Standards and Approved Food Additives*. In 1983 the first version of the Food Standards Code was published as the NHMRC *Model Food Standards Regulations*. However, adoption of these standards was still at the discretion of individual state legislatures, and uniformity was not always achieved (NHMRC, 1983).

In 1991 a commonwealth and state/territory agreement on uniform food standards laid the foundation for a single national statutory authority to replace the NHMRC in setting uniform national food standards. The outcome of this agreement was the formation of the National Food Authority (NFA), which was established by the National Food Authority Act 1991. Standards developed by the NFA required

Exhibit 5–1 State and Territory Legislation Implementing the National Health and Medical Research Council Model Food Act

State/Territory	Legislation
New South Wales	Food Act 1989
Victoria	Food Act 1984
Queensland	Food Act 1981
Western Australia	Health Act 1911
South Australia	Food Act 1985
Tasmania	Public Health Act 1962
Northern Territory	Food Act 1986
Australian Capital Territory	Food Act 1992

approval by a majority of commonwealth and state/territory health ministers. Thereafter such standards were incorporated into the Food Standards Code and adopted by reference, and without amendment, into all appropriate state and territory food legislation. Furthermore, states and territories agreed not to develop individual food standards outside of the agreement. In July 1996 a treaty between Australia and New Zealand resulted in New Zealand joining the Australian system. The National Food Authority Act 1991 was amended accordingly and became the Australia New Zealand Food Authority Act 1991, and the NFA was replaced by the Australia New Zealand Food Authority (ANZFA). The NFA and its successor, ANZFA, are administered through the health portfolio of the Australian Commonwealth Government.

FOOD REGULATION IN NEW ZEALAND

Food in New Zealand is regulated by the Food Act 1981 and its associated regulations, the Food Regulations 1984 and the Dietary Supplements Regulations 1985. Primary responsibility for food control rests with the Ministry of Health, with the Ministry of Agriculture and Fisheries, and with local authorities that have responsibility for various aspects.

Food standards are contained in the Food Regulations 1984. Proposals for the development or amendment of food standards were assessed by a Food Standards Committee, which represented consumers, the food industry, academia, and government. This committee recommended standards and variations to these to the Ministry of Health, which then coordinated further consultations on the standard within government and with particular interested parties. Where any specific issues or problems were identified, the draft standard was returned to the Food Standards Committee for further work. Once this iterative process of refinement was complete, the final standard was submitted to the Minister of Health for endorsement, and if agreed, passed to New Zealand Parliamentary Counsel for drafting into legislation and final approval by Government.

In 1993 it was decided that the Food Standards Committee arrangement should be replaced, and the committee did not meet again. In its place, an Officials' Committee on Food Administration was formed, and among other things was tasked with evaluating options for the future of food standards regulations. In 1994 this led to an agreement (Australia New Zealand [ANZ] Agreement) in principle to a joint standards system between Australia and New Zealand, and the subsequent establishment of that system in 1996.

JOINT ANZ STANDARDS

The aim of the 1996 food standards treaty between Australia and New Zealand was to create a Food Standards Code that would provide a system of food standards for general application across Australia and New Zealand. The system provides for food standards developed by the ANZFA and approved by a Ministerial Council, on which New Zealand now has a seat, to be adopted throughout Australia and New

Zealand by reference and without change (subject to the possibility of certain limited exceptions or variations in the case of New Zealand).

ANZFA and the Food Standards Code

The ANZFA is the realization of a partnership on food standards between the governments of the Commonwealth of Australia, the Australian States and Territories, and the Government of New Zealand. ANZFA's functions are summarized in Exhibit 5–2. ANZFA is an independent expert body established as a statutory authority under an Australian federal law: the ANZFA Act 1991. As a statutory authority, ANZFA fulfills a number of specific functions laid down in its legislation. The prime function of ANZFA is the development of food standards that form the Australia New Zealand Food Standards Code (ANZ Code).

ANZFA has a board that comprises a chair, a chief executive officer (CEO) (ex officio), and six part-time members, at least two of whom are nominated by New Zealand. Their expertise includes knowledge of the food industry, consumer affairs, food regulation, and food sciences. In addition, special purpose members may be co-opted onto the board for particular purposes.

To provide advice to the board, ANZFA employs approximately 70 full- and part-time staff, reporting to a chief executive. These include a high proportion of professionals with expertise in toxicology, microbiology, food technology, nutrition, food law, food communications, and food policy development.

In addition, ANZFA receives policy advice from the ANZFA Advisory Committee (ANZFAAC). ANZFAAC is the only standing committee of ANZFA. It is chaired by the ANZFA CEO and comprises senior officials nominated from each state and territory and New Zealand, as well as from specific agencies of the Australian federal government and the New Zealand government.

Recommendations made by ANZFA are considered by the Australia New Zealand Food Standards Council (ANZFSC). ANZFSC is made up of the health ministers of the Australian Commonwealth and of the states and territories, as well as the New Zealand Associate Minister of Health. ANZFSC may adopt, amend, or reject ANZFA's recommendations or refer them back for further consideration. In the event of a vote, standards are adopted by a simple majority.

Exhibit 5–2 Functions of the Australia New Zealand Food Authority

- The development of standards
- Coordination of food recalls
- Research and surveys
- Coordination of food surveillance
- Food safety education
- Development of assessment policies in relation to imported food
- Development of codes of practice for industry
- Advice to minister (Commonwealth)
- Incidental functions

Source: Data from the Australia New Zealand Food Authority.

Amending Food Standards

The Australian Food Standards Code (the Code), which will form the basis for the Joint ANZ Code, primarily standardizes food commodity categories. It also provides general standards for labeling, food additives, contaminants, and pesticide residues. However, by comparison with other national and international food standards regimes, it contains relatively few standards for specific food items.

The categories of food standardized in the Code include meat, fish, milk and dairy products, cereals and cereal products, fruits, and vegetables. When a single food commodity is not covered by a standard, it is considered to be a nonstandardized food. Food additives may not be added to single-ingredient nonstandardized foods. Foods that are mixtures of two or more foods, whether standardized or not, are normally addressed by application of a carryover principle, although the Code also gives a general permission for the use of many food additives in mixed foods.

In amending the Code, ANZFA is required to undertake an open and transparent process set out in the ANZFA Act 1991. Amendments to the Code can be initiated by anyone outside of ANZFA, in which case they are designated as applications, or they can be initiated by ANZFA itself, in which case they are designated as proposals. When considering applications, ANZFA must normally complete its process in 12 months, excluding time for the applicant to supply additional information. Proposals raised by ANZFA itself do not have a statutory time limit. The statutory process may be abbreviated by ANZFA in order to ensure that it is appropriate for minor or complex matters or, as a matter of urgency, to avoid one of the five objectives of food standards being compromised (see "Objectives of Food Standards").

The ANZFA processes also represent a significant change in the way decisions about food standards are made in that more focus is placed on outcomes. Previously, both the NHMRC in Australia and the New Zealand Food Standards Committee relied upon standing committees of experts and government representatives to make recommendations on food standards. In the ANZFA system, matters are handled and technical recommendations made by ad hoc project teams formed for that purpose. When ANZFA commences a project, a project manager is appointed, whether for the development of regulatory policy, to respond to an application, or to review a standard. Normally, the project manager will be an ANZFA staff member. The ad hoc project team is then assembled to bring relevant expertise to the project. For applications, teams are normally formed from ANZFA staff. However, for the development of the ANZ Code, membership of project teams is invited from relevant experts from government, industry, consumer groups, and appropriate professional bodies. Project managers and project teams are required to develop and then adhere to project plans that have clearly defined milestones and timelines to accord with the ANZFA statutory procedures.

Objectives of Food Standards

In order to provide a clear and agreed basis for the development of food standards, the ANZFA Act prescribes five objectives. These objectives are given in Exhibit 5–3.

Exhibit 5–3 The Objectives of Food Standards (in Priority Order)

> - The protection of public health and safety
> - The provision of adequate information relating to food to enable consumers to make informed choices and to prevent fraud and deception
> - The promotion of fair trading in food
> - The promotion of trade and commerce in the food industry
> - The promotion of consistency between domestic and international food standards where these are at variance
>
> *Source:* Data from the Australia New Zealand Food Authority.

The ANZ Agreement also requires that standards developed under the Australia New Zealand Food Standards System be in accordance with the following principles:

- protection of public health and safety, including provision of adequate information relating to food to enable consumers to make informed choices and to prevent fraud and deception
- facilitation of access to markets, including promotion of fair trading; trade and commerce; and consistency between the domestic food standards of the member countries and international food standards

The ANZ Agreement also requires that standards be consistent with the obligations of both member countries as members of WTO, consistent with domestic laws and regulations of both member countries, and in accordance with the following principles and practices:

- The standards must be based on the best available scientific data, including systematic application of public health risk analysis and risk management principles to the development of food standards.
- The standards must be of a generic nature where possible.
- The standards must be subject to the principles established by the Council of Australian Governments (COAG, 1995).

In addition, the ANZFA is also directed to take into account any issues raised in public submissions, the relevant New Zealand standards, and any other relevant matters. The ANZ Agreement contains common policy objectives for developing food standards and reaffirms the commitment to a common approach to a transparent, consultative, accountable, and timely standards-setting process.

Transitional Arrangements

Until the joint ANZ Code has been developed and implemented, the ANZ Agreement provides for certain transitional arrangements. On the date the ANZ Agreement was official, the Australian Code was adopted by New Zealand as a parallel and alternative food standards system to the existing New Zealand Food Regulations. Under the dual regulatory system producers, New Zealand manufacturers and importers may opt to comply with either the Australian Food Standards Code or the New Zealand Food Regulations.

In Australia, foods manufactured for domestic consumption must still in general comply with the Australian Food Standards Code although some specified products may comply with the New Zealand Food Regulations. However, foods imported into Australia from New Zealand may comply with either the Australian Food Standards Code or, with certain qualifications regarding pesticide residues, cadmium residues, and microbiological specifications, the New Zealand Food Regulations.

The Review of the Australian Food Standards Code

The ANZ Agreement recognizes that ANZ Code will be developed from the review of the Australian Food Standards Code. On the completion of the review, it will be formally recognized by both Australia and New Zealand on a date to be mutually determined. When that occurs, the transitional arrangements will end and the relevant New Zealand Food Regulations will cease to apply. In addition to the objectives of food standards discussed above, it is intended that the review will, where possible:

- Reduce the level of prescriptiveness of standards to facilitate innovation by allowing wider permission on the use of ingredients and additives, but with consideration of the possible increased need for consumer information.
- Develop standards that are easier to understand and make amendments more straightforward.
- Replace standards that regulate individual foods with standards that apply across all foods or a range of foods.
- Consider the possibility of industry codes of practice as an alternative to regulation.
- Facilitate harmonization of food standards between Australia and New Zealand.

In addition, the review is being conducted in accordance with competition policy principles approved by the Council of Australian Governments in 1995, which require the removal of unjustified obstacles to competition.

All review projects are treated as proposals to change the Code and are conducted according to the statutory requirements set out in the ANZFA Act, discussed above. These requirements include appropriate periods of public consultation. In addition, since September 1995 all ministerial councils, including the ANZFSC, have been required to conduct a regulatory impact analysis in the development of new or revised regulations, including food standards. Regulation impact statements (RIS) require that all reasonable regulatory options are considered and that there is a process of consultation with those likely to be affected by any proposed regulation. In New Zealand compliance cost statements are required. These statements are similar in their objectives to the Australian RIS.

Proposed Layout of Australia New Zealand Food Standards Code

Under the agreement between Australia and New Zealand, the ANZ Code may contain standards that relate to any of the following:

1. the safety of food, including its microbiological status
2. the composition of food, including the maximum or minimum amounts, where appropriate, of contaminants, residues, additives, or other substances that may be present in food
3. the method of sampling and testing the food to determine its composition and safety
4. the production, manufacture, or preparation of food
5. materials, containers, appliances, or utensils used in relation to food
6. the packaging, storage, carrying, delivery, or handling of food
7. any information about food including labeling, promotion, and advertising
8. such other matters affecting food as may affect the health of persons consuming food
9. the interpretation of other standards

Certain matters, however, remain individual to the two countries, such as maximum residue limits for agricultural and veterinary chemicals in food; food hygiene provisions, including requirements for food safety programs or other means of demonstrating the safety and compliance of foods; and export requirements relating to third country trade.

It was also agreed that all food standards contained in the Australian Food Standards Code as of the date of signing of the ANZ Agreement, other than Standard A14—Residues in Food (ANZFA, 1997a), would be included within the scope of the Australia New Zealand Food Standards System. ANZFA has, therefore, proposed the following layout for the ANZ Code, found in *Structure of the Australia New Zealand Food Standards Code* (ANZFA, 1998):

Part 1: Interpretation and Guide to Effective Use of the Code

Part 2: Generic Provisions [General definitions and preliminary provisions, and generic standards as follows:]
    2.1 Labeling [Labeling requirements applicable to all food. The policy relating to labeling will be explained. The requirements for food for which there is no specific definition or composition standard will be stated.]
    2.2 Food Additives [All provisions relating to all additives including technological function. It will also contain the specifications for additives.]
    2.3 Vitamins and Minerals
    2.4 Processing Aids
    2.5 Contaminants and Residues
    2.6 Food Contact Materials
    2.7 Analytical Methods and Sampling
    2.8 Microbiological Standards
    2.9 Foods Requiring Pre-Market Clearance

Part 3: Commodity Groups
    3.0 Foods in General

3.1   Dairy Products [excluding butter and dairy-fats]
3.2   Edible Fats and Fat Emulsions
3.3   Ice Cream and Edible Ices
3.4   Fruits and Vegetables [including fungi, nuts and seeds]
3.5   Confectionery
3.6   Cereals and Cereal Products
3.7   Breads and Bakery Products
3.8   Meat and Meat Products
3.9   Fish and Fish Products
3.10 Eggs and Egg Products
3.11 Sugars, Honey and Related Products
3.12 Salt, Herbs, Dried Spices, Condiments
3.13 Foods Intended for Particular Nutritional Uses
3.14 Nonalcoholic and Alcoholic Beverages
3.20 Mixed Foods

Part 4: Hygiene Standards [Australia only]

Part 5: Quick Reference Guide

WTO OBLIGATIONS

Both Australia and New Zealand are members of WTO and signatories to the Application of Sanitary and Phytosanitary Measures (SPS Agreement) and Agreement on Technical Barriers to Trade (TBT Agreement) (WTO, 1995). During the Uruguay Round of the GATT negotiations, which led to the establishment of the WTO, both Australia and New Zealand established their positions as strong supporters of trade liberalization. In addition, the ANZ Agreement on joint food standards explicitly requires ANZFA to ensure that food standards are consistent with the WTO obligations of both countries.

THE USE OF RISK ANALYSIS IN STANDARD SETTING

As stated in the Australia New Zealand Food Authority Act 1991 in its primary task of developing food standards, ANZFA must have regard first and foremost to protecting public health and safety. The assessment and management of health risks underlie the regulatory approach applied by ANZFA.

In general, there is a poor understanding in the community about food-related health risks and an unrealistic desire for a risk-free environment. ANZFA seeks to develop food standards that allow the sale of a wide range of food commodities. Some of these may carry an inherent, albeit low, level of risk and, in this regard, may be at odds with community expectations (ANZFA, 1996). In many cases, however, low levels of risk must be balanced against the health benefit of a nutritious and varied diet. An informed debate about acceptable levels of risk in relation to food can be achieved only through a greater awareness of risk assessment and management practices.

In order to foster community understanding of these issues, ANZFA published a policy paper entitled *Framework for the Assessment and Management of Food-Related Health Risks* in August 1996 (ANZFA, 1996). This paper details how ANZFA defines, assesses, and manages health risks in relation to the food supply in Australia and New Zealand. The concepts and procedures described are broadly consistent with those of other regulatory agencies and with principles established both by the Codex Alimentarius Commission (CAC), under the Joint Food and Agriculture Organization/World Health Organization (FAO/WHO) Food Standards Programme, and by the International Programme on Chemical Safety in cooperation with the Joint FAO/WHO Expert Committee on Food (CAC, 1996a, 1996b, 1997b; FAO/WHO, 1995, 1997a, 1997b).

THE REGULATION OF FOOD ADDITIVES

As a part of the review of the Food Standards Code, ANZFA is currently in the process of developing a joint ANZ general standard on food additives. As a prelude, ANZFA's predecessor, NFA, published a policy paper in Australia entitled *The Regulation of Food Additives* in March 1996 (NFA, 1996); the policies in this paper were recognized by ANZFA in its proposal for a joint ANZ standard for food additives (ANZFA, 1997b). The NFA policy paper recognized that the ANZFA in considering approval of the use of food additives will continue to apply the Codex guidelines on the use of food additives. It also will, when considering an application for new additives and processing aids or variations in use of approved additives and processing aids, consider as far as practicable other possible uses and the need to vary other standards not covered by the application. Therefore, the considerations for accepting a food additive are

- It poses no unacceptable risk to health when used in amounts up to the approved limits even after a lifetime of consumption.
- There is a demonstrable need for the substance and that it fulfills established criteria for technological function that, in effect, provide benefits to consumers.
- The substance is used in any food only up to the level that achieves the technological function, regardless of the fact that higher levels might pose no threat to health. This is the concept of limited use according to good manufacturing practice.

ANZFA has recognized the policies set out in this paper in its proposal P150 for a joint ANZ standard for food additives.

ANZFA considers that approval of an additive is contingent on assurance that the accepted safe level of intake will not be exceeded even by consumers of large amounts of those foods in which the additive is to be permitted. Decisions about the appropriate levels of prescription necessary for each food additive within the standard are made *inter alia* with regard to the toxicological profile of the additive, to the likely limits of exposure from all expected uses, and to the potential for fraud and deception arising from the misuse of the additive. In applying the policy of regulating additives in generic standards, wherever practicable, ANZFA attaches a high level of importance to risk analysis (risk assessment and risk management). A move to generic standards results in a greater need to be confident about the safety aspects of particular additives and, in particular, the scientific basis for the acceptable level of exposure.

In contrast, decisions about which additives are justified are made on the basis of the nature of the foods and the technological function being achieved under normal conditions, rather than from the individual needs of specific products or the performance of individual additives. For example, a decision regarding whether a preservative is necessary in a soft drink is based on the likelihood that the food will spoil when packaged in plastic bottles under conditions of good manufacturing practice (GMP), not on whether the product could be hot filled into glass bottles, thereby obviating the need for a preservative. Which preservatives are then permitted is based upon an assessment that exposure (from all permitted uses) will not exceed recognized safe limits (e.g., the acceptable daily intakes [ADIs]) where these have been established, and the use of the specific additives will not be deceptive or misleading.

Defining the concept of a food additive has also been a matter of some discussion in Australia and New Zealand. Currently, regulation is achieved by a general prohibition on additive use unless a specific permission is given via positive list in a standard. However, such an approach is highly prescriptive and does not always facilitate innovation. From a regulatory perspective, a greater reliance on a definition may be seen to be important in the implementation of a less onerous system of premarket clearance of new additives and new categories of use for existing additives.

In developing a food additive standard, an agreed definition of food additives was valued. However, from an enforcement perspective, the positive list was seen to be a more effective approach. To combine these two approaches, a working definition of a food additive was developed, which drew upon the European Union (EU) definition, and linked this to a clearly established set of technological functions:

> A food additive is any substance not normally consumed as a food in itself and not normally used as a characteristic component of food which is intentionally added to a food to achieve one or more of the technological functions specified and which results, or may be reasonably expected to result, in it or its by-products becoming a component of such foods.

In effect, this solution draws upon both regulatory approaches described above, with additives in the holistic sense being prohibited, but with broad exceptions being granted for enumerated substances and uses based upon identified technological functions. The ANZFA policy paper proposes that this approach is sufficiently flexible to tailor the permissions granted for each additive, with some substances being granted broad permission for use in a variety of functional uses, and others being limited to specified amounts in particular foods for a specified function. ANZFA also foreshadowed that consideration would be given to the development of a set of guidelines to aid in the identification of food additives to support the operation of the new standard.

The ANZFA policy paper on the regulation of food additives also noted that confusion is often expressed about the differences between food additives and processing aids and that many of the existing definitions of additives do not distinguish clearly between them. This confusion is compounded by the fact that the assessment of use and safety of processing aids is carried out on a similar basis to that for food additives. ANZFA concluded that while processing aids may be seen as a subgroup of additives from a technical perspective, they are distinguished by the fact that their function is performed during manufacture, processing, transport, and storage of a food, rather than when it is prepared for consumption.

Intent of Food Additive Standards

The ANZFA food additives policy paper describes the principal intent of food additive standards as being to ensure that intake of food additives from the food supply does not present a risk to public health and safety. This is achieved by establishing maximum permitted levels for food additives in relevant foods where a potential risk to public health and safety may be identified. The levels of additives that may be added should be established on the basis of a risk analysis that takes into account appropriate measures of safety (such as the ADI), levels of additives required to achieve relevant technological functions, and estimated daily intakes from all relevant foods.

The policy paper proposes a second intent of food additive standards as being to ensure that consumers are not exposed unnecessarily to high levels of food additives. The establishment of limits is therefore seen to be appropriate where there is a risk of consumers being misled by the inappropriate use or overuse of a food additive in particular foods or categories of foods.

Thereafter it is proposed that an ANZ food additive standard should facilitate both the consumers' desires to exercise choice in selecting foods and the food industry's need to be innovative. Both of these goals are achieved by applying the minimum restriction on use consistent with good manufacturing practice (GMP).

Technological Need

The policy paper also addresses the issue of technological need in the use of food additives, noting that use of a food additive should be linked to a technological function. It is generally assumed that on the basis of information from the food industry and the community, regulatory agencies determine those functions that are technologically justified. In practice, food additive standards allow the use of many additives often by functional class in broad food categories rather than on a case-by-case basis. This effectively leaves the decision to industry and consumers.

Such standards, which permit the use of food additives other than those that operate on a recipe-by-recipe basis, may potentially allow the addition of additives to individual food products in which their use cannot be technologically justified.

Furthermore, current Australian (and New Zealand) standards do not elaborate criteria upon which technological justification can be assessed by food manufacturers or consumers. Consequently, within the boundaries of a permitted use, abuse of that additive may still occur. Therefore, ANZFA concluded that the process of deciding which technological functions are appropriate in a particular food can be undertaken in isolation from consideration of individually named food additives. Once it has been decided that a function (e.g., the need for an emulsifier or an antioxidant) is justified on technical grounds, a list of additives that can perform this function may be drawn up. Safety considerations and the potential for fraud and deception may limit the additives that are permitted as well the levels that may be used. However, thereafter, choices about which additive to select from the list should depend upon manufacturing processes and techniques and are, therefore, properly beyond the scope of a food additive standard.

ANZFA also references the CAC document *General Principles for the Use of Food Additives* (CAC, 1992), which provides guidelines on the justification of technological functions:

> The use of food additives is justified only where they serve one or more of the purposes set out from (a) to (d) and only where these purposes cannot be achieved by other means which are economically and technologically practicable and do not present a hazard to the health of the consumer:
>
> a. to preserve the nutritional quality of the food; an intentional reduction in the nutritional quality of a food would be justified in the circumstances dealt with in subparagraph (b) and also in other circumstances where food does not constitute a significant item in a normal diet;
>
> b. to provide necessary ingredients or constituents for foods manufactured for groups of consumers having special dietary needs;
>
> c. to enhance the keeping quality or stability of a food or improve its organoleptic properties, provided that this does not change the nature, substance or quality of the food so as to deceive the consumer; and
>
> d. to provide aids in the manufacture, processing, preparation, treatment, packing, transport or storage of food, provided that the additive is not used to disguise the effects of the use of faulty raw materials or of undesirable (including unhygienic) practices or techniques during the course of any of these activities.

Good Manufacturing Practice

ANZFA also recognizes that the use of a food additive should not be a substitute for GMP, and food additives should always be used in accordance with GMP. Permission to use a particular additive in a food or class of food should be qualified by a general requirement that any individual use must be consistent with GMP. This concept is not currently addressed in the Australian Food Standards Code or the New Zealand Food Regulations, and many additives are permitted in both countries without limit.

Although well-established in Codex standards (CAC, 1997a) and the EU directive (as *quantum satis;* EC, 1995), the introduction of a reference to GMP in a joint ANZ food standard raises a new complexity for enforcement authorities, particularly in establishing the appropriate level of an additive necessary to achieve a desired function in a specific food. ANZFA considers that the justification for the use of unusually high levels of an additive should be the responsibility of the manufacturer. The elaboration of clear guidelines for GMP in association with a revised standard discourages the unnecessary use of additives. In this regard, ANZFA cites the CAC's *Procedural Manual* (CAC, 1995a), which sets out the following relevant criteria for use in assessing compliance with GMP:

- the quantity of additive added to food shall be limited to the lowest possible level necessary to accomplish its desired effect;
- the quantity of the additive that becomes a component of food as a result of its use in the manufacture, processing or packaging of a food, and which is not intended to accomplish any physical or other technical effect in the finished food itself, is reduced to the extent reasonably possible; and
- the additive is prepared and handled in the same way as a food ingredient.

A Proposal for a Joint ANZ General Standard for Food Additives

ANZ started the statutory process of developing a new ANZ general standard for food additives by the release of Proposal P150 for a standard in March 1997 (ANZFA, 1997b). It is expected that this process will be completed with the implementation of the new standard in the second half of 1999. The proposal draws on the NFA policy paper (1996) as well as the progress made by the Codex Committee on Food Additives and Contaminants toward the development of a General Standard for Food Additives (CAC, 1997a). The proposed draft standard does not define a food additive as such although it proposes the working definition discussed above in an interpretation note.

All food additives permitted currently in Australia and New Zealand (as well as a number permitted in the EU, Canada, and the United States for which safety data are in the public domain) were considered in the preparation of the proposed draft standard. Those additives that on an initial screening of available toxicology and intake data were found to be suitable for continuing food use were proposed for inclusion in the draft standard. Many of these additives, particularly those with general, rather than highly specific functions, are already permitted and used widely in food in both countries. In many cases, no maximum levels of use are set at present.

The proposed draft standard applies greater consistency in the way many of these additives are permitted and introduces the concept of GMP as the effective limit on excessive use of these additives. Colors were addressed separately within the draft standard to reflect community concerns about the use of these specific additives. Flavorings were not addressed in detail at this stage, and the draft merely included flavorings by reference to Standard A6—Flavorings and Flavor Enhancers (ANZFA, 1997c). It is proposed that the contents of Standard A6 will be addressed in detail in a later review.

To facilitate the development of the proposed draft standard the additives were arranged into five groupings:

Group 1. Miscellaneous additives, currently permitted extensively in Australia and/or New Zealand, for which numerical ADIs are currently considered not necessary because of a lack of observed toxicity or which have numerical ADIs that are unlikely to be approached from all technically justified uses. These additives were listed in a schedule (Schedule 2) to the standard. In general, the technological functions achieved through the use of these additives are organoleptic, and within broad limits, the justification and appropriate level of their use are subject to individual preferences. Group 1 additives were, therefore, proposed in accordance with GMP in foods except fresh and unprocessed foods or where the general presence of food additives would not be reasonably expected.

Group 2. Colors for which ADIs have been deemed unnecessary on account of their lack of observed toxicity or which have numerical ADIs that are unlikely to be approached from all technically justified uses. The justification for use of color additives in processed foods is largely a matter of individual preference and essentially similar to that set out for Group 1 additives. These additives were listed in a separate schedule (Schedule 3) to the proposed draft standard and permitted in accordance with GMP in specified processed foods.

Group 3. Colors that have numerical ADIs that are sufficiently high to enable their inclusion at a technologically useful level in all processed foods when tested on a dietary budget model. These additives are listed in a separate schedule (Schedule 4) to the proposed draft standards and permitted subject to defined limits in specified processed foods.

Group 4. Food additives that have specific uses for which they can be generally considered as safe. This includes some individual additives from groups 1 or 2 in specific foods such as unprocessed foods (e.g., waxed fruit and vegetables) that are not permitted to contain additives in general. These additives were listed in the main schedule (Schedule 1) to the standard under specific categories of food and limited both by defined limits and/or GMP.

Group 5. Additives with numerical ADIs, where preliminary estimates of potential intake indicate could be exceeded by unrestricted use, and individual coloring additives from Group 3 in specific foods where Group 3 additives are not generally permitted. These additives are permitted, subject to defined limits by individual food category in Schedule 1 to Standard A6.

In grouping the additives, consideration was given to the results of safety evaluations undertaken by ANZFA as well as by other relevant agencies including the National Health and Medical Research Council, the Joint FAO/WHO Expert Committee on Food Additives, the Scientific Committee for Food of the European Commission, Health and Welfare Canada, and the U.S. Food and Drug Administration.

The Australia New Zealand Food Identification System

A food identification system used in the proposed draft Australia New Zealand general standard for food additives was developed from the food classification system developed by the Confederation of Food and Drink Industries of the European Community. This system has also been adopted, with variations, by the Codex Committee on Food Additives and Contaminants for use in the Codex General Standard on Food Additives.

The Australia New Zealand Food Identification System (ANZFIS) is designed to allow the specific identification of individual foods or classes of foods to which additives may be added or from which additives are to be expressly excluded. It is intended that this will increase clarity in the use of food additives and enable greater confidence in estimating exposure to individual additives. The food classification system has also been useful for determining the way in which commodity and product standards will be grouped in the revised ANZ Code.

THE REGULATION OF CONTAMINANTS AND OTHER RESTRICTED SUBSTANCES

Definition of a Contaminant

Food contaminants are generally considered to be those substances present in food at levels that serve no technological function and whose presence may lead to adverse health effects (ANZFA, 1997d). In the majority of cases, contaminants serve no nutritional function, although some, such as copper, selenium, and zinc, are essential micronutrients that may have an adverse health effect at high levels of consumption. There is currently no Australian or New Zealand definition of a contaminant. The CAC defines (CAC, 1995b) a contaminant as:

> Any substance not intentionally added to food, which is present in such food as a result of production (including operations carried out in crop husbandry, animal husbandry and veterinary medicine), manufacture, processing, preparation, treatment, packing, packaging, transport or holding of such food or as a result of environmental contamination. The term does not include insect fragments, rodent hairs and other extraneous matter.

Within the scope of this definition, the potential range of food contaminants is very large, but could be interpreted as restricted to synthetic chemicals, metals, mycotoxins, and bacterial toxins. Synthetic chemicals may include any industrial chemical or environmental contaminant. Because the definition stipulates "not intentionally added to food," the definition excludes agricultural and veterinary chemicals, food additives, and processing aids. However, certain pesticides, such as DDT, that are no longer intentionally added to crops, but are still found in foods, may also be regarded as contaminants.

Chemicals that occur naturally in foods (e.g., plant and marine toxins) are not, strictly speaking, contaminants since they are inherent components of the plant or marine animal. It is debatable whether they are included under the Codex definition of a contaminant although CAC standards have been proposed for a number of these substances. The Codex classification of contaminants is provided in Exhibit 5–4.

A number of naturally occurring toxins are currently controlled under the Australian Food Standards Code and the New Zealand Food Regulations. The Australian Food Standards Code also includes an extensive list of prohibited botanicals that are also unlikely to be included within the Codex definition. In order to accommodate these groups of substances, a broader definition of a contaminant will be needed or the title of the standard changed to include reference to these restricted substances.

Controlling the Level of Contaminants in Food

Contamination of food with potentially hazardous substances can occur at all stages of food production (ANZFA, 1997d). In many cases, the level of potential contamination with substances from environmental sources is self-limiting because of current manufacturing practices and adherence to agricultural or industrial best practices. A generally recognized benchmark for control of contaminants is that levels in food should be as low as reasonably achievable. This is normally taken to mean concentration of a substance that cannot be eliminated from a food without involving the discarding of that food altogether.

Action to ensure that contamination of food is as low as reasonably achievable may be taken in a variety of ways, including the development of control measures at the source of contamination, the enforcement of both good agricultural practice and GMP and, if necessary, the establishment in food standards of maximum permitted concentrations (MPCs) of contaminants in particular commodities. The MPC for a contaminant is the maximum concentration of a substance permitted in a food commodity. MPCs are the legal limits enforced through food standards. The decision about whether control is necessary and appropriate is based on consideration of the following points:

- the potential for a human health risk posed by the contaminant
- the nature and severity of the potential health risk, particularly in susceptible population groups

Exhibit 5–4 Codex Classification of Contaminants

- Metals, metalloids, and their compounds
- Other elements and inorganic compounds
- Hallogenated organic compounds
- Other organic compounds
- Mycotoxins
- Other microbial and food-processing–related toxins
- Phytotoxins and other inherent naturally occurring toxins
- Radioactive isotopes

Source: Data from *Codex Alimentarius Commission Procedural Manual*, 9th Ed., © 1995, Codex Alimentarius Commission.

- the frequency with which contamination of a particular food occurs
- the importance of the food in the total diet (especially staple foods and human milk) and identification of the most exposed consumer groups
- the feasibility of measuring the level of the contaminant in a reliable manner in an adequate number of food samples
- whether the contaminant is also a micronutrient (e.g., copper); the potential nutritional effect of controlling the contaminant must be taken into consideration, as well as the potential for micronutrient interaction

These points are consistent with the considerations expressed in the 1992 Report of the Joint UNEP/FAO/WHO Food Contamination Monitoring and Assessment Programme (UNEP/FAO/WHO, 1992).

If the source of the contaminant can be identified, consideration should be given to whether entry of the contaminant into the food chain can be prevented at this point. For example, polychlorinated biphenyl contamination of food may be controlled by reducing its environmental levels. Similarly, for mycotoxins, improvements in food production, handling, and processing may prevent or reduce aflatoxin contamination of foods. For some contaminants (e.g., ergot in grains), decontamination processes may be available to reduce the level of contamination before processing. If appropriate, these procedures may be incorporated into relevant industry food safety plans. Food standards, in the form of maximum permitted concentrations of contaminants, are generally used when other mechanisms of control are considered insufficient or inadequate to safeguard the health of consumers.

Principles for Establishing Maximum Permitted Concentrations of Contaminants

As part of a review of MPCs for cadmium, principles for controlling the level of contaminants in food and setting MPCs were adopted by ANZFA in 1997 and endorsed by the Council of State, Territory, and New Zealand Health Ministers (the Australia New Zealand Food Standards Council). ANZFA evaluated the establishment of MPCs for cadmium using a principle that would apply to any consideration of food contaminants and is consistent with that adopted by the CAC. Namely, contaminant levels in food should be safe and as low as reasonably achievable. The maximum permitted concentrations for food should reflect this principle, which is based on the premise that contaminants have no intended function in food and their associated health risks may not yet be fully understood.

ANZFA applies the following additional principles when evaluating the establishment of MPCs for contaminants. These are secondary to the broader section 10 objectives of the Australia New Zealand Food Authority Act 1991:

1. An MPC will be established only where it serves an effective risk management function; and MPCs will be set for:
   a. all primary commodities (described using Codex food commodity groupings) that provide, or may potentially provide, a significant contribution

　　　　to the total dietary contaminant intake, as indicated by dietary exposure assessments
　　b.　nominated processed foods where the setting of an MPC for the primary commodity is judged to be ineffective
2.　An MPC will be set at a level that is consistent with public health and safety as determined by an appropriate risk assessment procedure based on dietary modeling and which is reasonably achievable from sound primary production and natural resource management practices. Australian and New Zealand data will normally be used for this purpose.
3.　In setting an MPC, consideration will be given to Australia and New Zealand's international trade obligations under the SPS and TBT Agreements.
4.　Other measures that might be used to reduce contaminant levels in the food supply and consequent dietary intakes include improving primary commodity production practices and developing appropriate education programs for population groups with potential for high exposure to particular contaminants.

Factors To Be Considered in Establishing an MPC

For the majority of potential contaminants, setting an MPC in food may not be considered necessary or appropriate since there are many other effective mechanisms for limiting the contamination of food, both for agricultural commodities and for processed foods. There are, for example, zoning restrictions on the use of land for agricultural purposes near major industrial complexes or the siting of food manufacturing premises. Similarly, good agricultural practices and good manufacturing practices significantly reduce the potential for food contamination.

In assessing whether an MPC is necessary for a particular contaminant, the effectiveness of existing controls for minimizing contamination and the potential for these controls to fail both need to be considered. Risk assessment procedures can be used to consider whether the potential exposure to a contaminant may be a cause for public health concern. It is particularly useful to assess the risk associated with residual levels of contaminants that occur despite existing controls.

The principal criteria for establishing an MPC for a contaminant in the revised standard will be: (1) demonstration of a potential adverse health effect at the anticipated levels of total dietary exposure, or, where necessary, (2) demonstration of the need for a control mechanism to ensure that the level of the contaminant in food is as low as reasonably achievable.

In order to ensure that the contaminant levels achieved through sound production and natural resource management practices do not result in adverse health, it is necessary to determine the total level of dietary exposure to the contaminant from each individual food commodity and the level of dietary exposure for the high consumer of a particular food commodity. Dietary modeling procedures can be used for this purpose.

In the majority of cases, initial contamination occurs in the raw agricultural produce rather than in processed foods and, therefore, MPCs are normally set for raw agricultural products only. Further processing, unless it involves dehydration or separation of food components, would be expected to lead to a reduction in the level of contamination. The maximum level allowed in raw products should be set on the basis of a knowledge of the total consumption of food containing the raw commodity. There may be a need to have a dehydration clause in the standard that allows a readjustment of the permitted level before comparison of contaminant levels to the established standard, because some foods are consumed in a dehydrated state.

For each food commodity, MPCs for a particular contaminant are generally established at the upper end or above the range of contaminant levels normally found in the food commodity, provided that this level is consistent with public health and safety. The MPC is established on the basis that sound production methods and sound natural resource management practices have been used throughout the food production system. In order to establish that a proposed MPC will not lead to adverse health outcomes, it is compared to a theoretical upper-level MPC based on health criteria. This can be determined on the basis of the level of the contaminant that would need to be in a commodity in order for a high consumer of that particular commodity to exceed the provisional tolerable weekly intake.

In most cases, an MPC based on health criteria (even for the high consumer) will be considerably greater than that determined from the range of contaminant levels normally found in the food commodity. If there were a situation where the proposed MPC could lead to potential health problems for the high consumer (as determined from dietary modeling), a lower MPC might be necessary. In this unlikely situation, additional steps would be necessary to reduce the overall level of contaminant in that particular food commodity.

While food standards are the risk management instrument within the food regulations for controlling contamination, they must operate within a broader risk management strategy in order to achieve and maintain a low health risk for all consumers. Regulations operate at all government levels and in a number of agencies to encourage practices that will lead to minimization of food contamination. These include waste management/disposal programs, water quality control programs, industrial zoning regulations, and environmental safeguards.

Two of the more direct complementary strategies for control of contamination are: (1) improvements in primary food production practices, in order to minimize contamination of the primary commodity; and (2) appropriate education programs for population groups with particular dietary habits that lead to a potential for high exposure to particular contaminants.

CONCLUSION

Food standards for additives and contaminants in Australia and New Zealand are developed within a framework that seeks to ensure consumer safety while at the same time allowing for industry innovation and encouraging international trade. In order to achieve these goals, an open and consultative process has been established that allows consideration of all viewpoints. The principles for establishing standards for additives and contaminants are based on the broader objectives of food standards as defined in the Australia New Zealand Food Authority Act 1991.

REFERENCES

Australia New Zealand Food Authority. (1996). *Framework for the Assessment and Management of Food-Related Health Risks*. Canberra, Australia: Australia New Zealand Food Authority.

Australia New Zealand Food Authority. (1997a). 1997 Food Standards Code, Commonwealth of Australia. Canberra, Australia: Australia New Zealand Food Authority.

Australia New Zealand Food Authority. (1997b). *Review of the Australian Food Standards Code: Proposal P150. A Proposal for a Joint Australian New Zealand Standard on Food Additives*. Canberra, Australia: Australia New Zealand Food Authority.

Australia New Zealand Food Authority. (1997c). Standards A6—Flavorings and Flavor Enhancers, 1997 Food Standards Code, Commonwealth of Australia. Canberra, Australia: Australia New Zealand Food Authority.

Australia New Zealand Food Authority. (1997d). *The Regulation of Contaminants and Other Restricted Substances in Food*. Canberra, Australia: Australia New Zealand Food Authority.

Australia New Zealand Food Authority. (1998). *Structure of the Australia New Zealand Food Standards Code*. Canberra, Australia: Australia New Zealand Food Authority.

Codex Alimentarius Commission. (1992). General principles for the use of food additives. In *Codex Alimentarius*, Vol. I (General Requirements), pp. 49–51, 2nd ed. Rome: Codex Alimentarius Commission.

Codex Alimentarius Commission. (1995a). *Codex Alimentarius Commission Procedural Manual*, 9th ed. Rome: Codex Alimentarius Commission.

Codex Alimentarius Commission. (1995b). *Preamble to the Codex General Standard for Contaminants and Toxins*. Codex Alimentarius Commission, 21st Session, ALINORM 95/12A, Appendix VI. Rome: Codex Alimentarius Commission.

Codex Alimentarius Commission. (1996a). *Terms and Definitions Used in Risk Assessment*. Doc. CX/EXEC. 96/43/6, Annex 1. Rome: Codex Alimentarius Commission.

Codex Alimentarius Commission. (1996b). *Codex Risk Assessment and Management Procedures: Methods To Ensure Public Safety While Developing the Codex General Standard for Contaminants and Toxins in Food*. CX/FAC 96/15. Rome: Codex Alimentarius Commission.

Codex Alimentarius Commission. (1997a). Preamble to the General Standard for Food Additives and schedule of additives permitted for use in food in general unless otherwise specified. In *General Standard for Food Additives*. Codex Alimentarius Commission, 22nd Session, Geneva, ALINORM 97/12A. Rome: Codex Alimentarius Commission.

Codex Alimentarius Commission. (1997b). *The Application of Risk Analysis Principles in Codex*. Codex Alimentarius Commission, 22nd Session, Geneva, ALINORM 97/9. Rome: Codex Alimentarius Commission.

Council of Australian Governments. (1995). *Principles and Guidelines for National Standard Setting and Regulatory Action by Ministerial Councils and Standard Setting Bodies*. April 1995. Canberra: Council of Australian Governments.

European Community. (1995). European Parliament and Council directive 95/2/EC of 20 February 1995 on food additives other than colours and sweeteners. *Off J Eur Communities* L 61, 1–40.

Food and Agriculture Organization/World Health Organization. (1995). *Application of Risk Analysis to Food Standards Issues*. Report of the Joint FAO/WHO Expert Consultation. WHO/FNU/FOS/95.3. Geneva, Switzerland: World Health Organization.

Food and Agriculture Organization/World Health Organization. (1997a). *Risk Management and Food Safety*. Report of a Joint FAO/WHO Consultation. Rome: Food and Agriculture Organization.

Food and Agriculture Organization/World Health Organization. (1997b). *Joint FAO/WHO Consultation on Food Consumption and Exposure Assessment of Chemicals*. Geneva, Switzerland, February 10–14, 1997. Geneva, Switzerland: World Health Organization.

National Food Authority. (1996). *The Regulation of Food Additives*. Canberra, Australia: National Food Authority.

National Health and Medical Research Council. (1982). *Approved Food Standards and Approved Food Additives*. Canberra, Australia: Australian Government Printing Service.

National Health and Medical Research Council. (1983). *Model Food Legislation*. Canberra, Australia: National Health and Medical Research Council, Australian Government Printing Service.

United Nations Environment Programme/Food and Agriculture Organization/World Health Organization. (1992). *Assessment of Dietary Intake of Chemical Contaminants*. Nairobi, Kenya: United Nations Environment Programme.

World Trade Organization. (1995). *The Results of the Uruguay Round of Multilateral Trade Negotiations: The Legal Texts*. Geneva, Switzerland: World Trade Organization.

CHAPTER 6

Southern Common Market Standards

Maria Cecilia de Figueiredo Toledo

INTRODUCTION

Founded to achieve economic cooperation among four Latin American countries, the Southern Common Market (MERCOSUR) comprises nearly 190 million people living in an area larger than the total surface of the European continent, covering more than 12×10^6 km^2. Setting as one of its principal objectives the establishment of a single common market in Latin America, MERCOSUR was created in 1991 under the Treaty of Asunción (SICE, 1999a), signed by Argentina, Brazil, Paraguay, and Uruguay.

Although MERCOSUR is not yet fully implemented, in 1997 intra-MERCOSUR trade totaled $19.7 billion, representing an increase of 386.5 percent over the volume of 1991 (MRE, 1999a). In 1997 the total gross domestic product of these four nations was almost $1.23 trillion (MRE, 1999b). MERCOSUR is part of the world globalization process and is a unique opportunity for economic growth while ensuring the availability of safe and nutritious food to consumers. Since food is one of the most important commodities in international trade, the harmonization of national food standards has been considered a priority.

MERCOSUR countries have different laws governing food that are based, among other things, on historical, cultural, and economic factors. Some regulations are complex and contain many controls while others are less developed and lack basic requirements. As a consequence, the development and adoption of a common legislation has been and will continue to be long and difficult. Although not immediately apparent, the difficulty in achieving consensus within MERCOSUR is not unlike what the members of the European Union (EU) experienced. Despite a common wish to harmonize trade, during negotiations, member countries want to

---

Acknowledgments: The author thanks Dr. Leo Bick, technical director of ABIA, for his helpful comments and suggestions on this chapter.

preserve their own regulations. On many occasions, decisions are not made solely on the basis of scientific data.

Currently, food harmonization within MERCOSUR has been reached for issues where technical standards may represent serious trade barriers. These decisions were based on the EU experience as well as on Codex Alimentarius Commission (CAC) guidelines and recommendations. MERCOSUR member countries are conscious of their commitments to the World Trade Organization (WTO) and plan to adjust their national laws to conform to this agreement.

This chapter discusses the MERCOSUR single market with emphasis on its food law harmonization program. References to structure and scope and main provisions of the Food Commission are included.

HISTORICAL BACKGROUND

Latin America took its first steps toward regional integration with the Latin American Free Trade Association (ALALC), in a treaty signed in 1960 (MRE, 1999c). This treaty was designed to create a free trade zone, by means of periodic and selective negotiations between its member countries. In 1980, the Latin American Integration Association (ALADI), which established an economic preference zone in place of the free trade zone, replaced ALALC (ALADI, 1999).

In 1986 under the ALADI, Brazil and Argentina signed several commercial protocols that represented their first concrete steps toward aligning controls in the two countries. This alignment officially started in 1985 under the Declaration of Iguaçu (MRE, 1999d). To supplement and improve on their former agreements, in 1988 Brazil and Argentina signed a Treaty for Integration, Cooperation, and Development that set the stage for a common market between the two countries within 10 years, with the gradual elimination of all tariff barriers and harmonization of the macroeconomic policies of both nations. It was further established that this agreement would be opened to all other Latin American countries. After the accession of Paraguay and Uruguay, these two countries as well as Brazil and Argentina signed a new treaty (SICE, 1999a) on March 26, 1991, in Asunción, Paraguay. This provided for the creation of a common market among the four participants, to be known as the Southern Common Market or MERCOSUR. The integration process had a transition period that ended in December 31, 1994, during which Bolivia, Chile, Venezuela, Colombia, and Peru showed interest in joining MERCOSUR.

In June 1996, after a long period of negotiation, the presidents of the MERCOSUR member countries and the president of Chile signed an agreement to create a free trade zone among the five countries within a period of eight years. Subsequently, a similar agreement was signed with Bolivia to gradually reduce its trade tariffs among MERCOSUR nations within the period of 10 years.

OBJECTIVES OF MERCOSUR

The objectives of MERCOSUR, as stated by the Treaty of Asunción (SICE, 1999a), include:

- free transit of goods, services, and production factors between the member countries; establishment of a common external tariff (TEC); and the adoption of a common trade policy with regard to nonmember countries
- coordination of macroeconomic and sectorial policies of member countries in order to ensure free competition
- commitment by the member countries to harmonize their laws to allow for the strengthening of the integration process

The Treaty of Asunción is based on reciprocal rights and obligations between the member countries. In addition to the reciprocity principle, the Treaty of Asunción also contains provisions ensuring equitable trade conditions by using national laws to inhibit imports whose prices are influenced by subsidies, dumping, or unfair practices.

The evolution of the European Community into the EU has proved an interesting example of how legislation was harmonized between the member countries, and MERCOSUR countries have observed and learned from this experience.

INSTITUTIONAL STRUCTURE

The Treaty of Asunción and the Ouro Preto Protocol (SICE, 1999b) established the basis for the institutional structure of MERCOSUR (Figure 6–1), thereby creating the Common Market Council (CMC), the Common Market Group (GMC), the MERCOSUR Trade Commission (CCM), the Joint Parliamentary Commission (CPC), the Socioeconomic Advisory Forum (FCES), and the MERCOSUR Administrative Office (SAM).

The Common Market Council

The CMC, composed of the Ministries of Foreign Affairs and the Ministries of Economy (or the equivalent) of all four countries, is the highest level agency of MERCOSUR with authority to conduct its policy and responsibility for compliance with the objectives and time frames set forth in the Treaty of Asunción. Member countries preside over the CMC in rotating alphabetical order, for six-month periods. CMC members meet whenever necessary, at least once a semester, with the participation of the member countries' presidents. Included among its functions are the negotiation and the signature of agreements with third countries, groups of countries, and international organizations, as well as the creation, modification, or extinction of MERCOSUR bodies. All decisions taken by the CMC are mandatory in the member countries.

The Common Market Group

The GMC is the executive body of MERCOSUR and is coordinated by the Ministries of Foreign Affairs of the member countries. Its basic duties are to ensure compliance with the Treaty of Asunción and to take resolutions required for implementation of the decisions made by the CMC. It has the authority to organize,

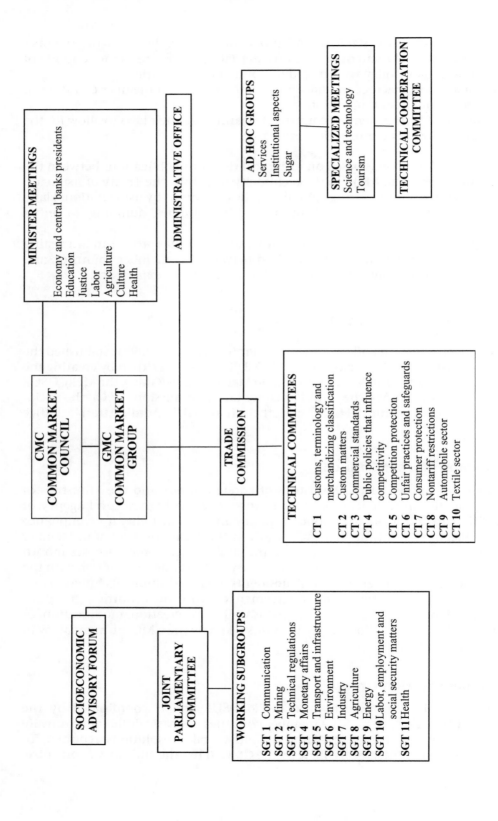

Figure 6–1 Institutional Structure of MERCOSUR

coordinate, and supervise work subgroups and to call special meetings to deal with issues of interest. Its members represent the Ministry of Foreign Affairs, the Ministry of the Economy (or the equivalent), and the president of the Central Bank. The GMC meets ordinarily at least once every quarter in the member countries, in rotating alphabetical order. GMC decisions are made by consensus, with the representation of all member countries. The official MERCOSUR languages are Portuguese and Spanish.

Directly subordinated to the GMC are the work subgroups, whose responsibilities include studies on specific MERCOSUR subjects and the preparation of recommendations for its consideration and approval as MERCOSUR resolutions. At present, the work subgroups are SGT-1, Communication; SGT-2, Mining; SGT-3, Technical Regulations; SGT-4, Monetary Affairs; SGT-5, Transport and Infrastructure; SGT-6, Environment; SGT-7, Industry; SGT-8, Agriculture; SGT-9, Energy; SGT-10, Labor, Employment and Social Security Matters; and SGT-11, Health.

The meetings of the work subgroups are held quarterly in the member countries, alternating every six months in alphabetical order. Their activities are carried out with the participation of representatives from the official and private sectors of each member state. Industry representatives are directly involved in the review stages of the production, distribution, or consumption process, for the products that fall within the scope of the subgroup's activities. The decision-making stage is reserved exclusively for official representatives.

The MERCOSUR Trade Commission

The CCM, coordinated by the Ministries of Foreign Affairs, assists the GMC by applying the instruments of common trade policy agreed by the member countries for the operation of the customs unification, as well as by following up and revising issues related to common trade policies within MERCOSUR and with other countries. The CCM has four members and four alternates. It meets at least once a month or whenever asked by the GMC or any of the member countries.

The CCM also considers demands of member countries or private parties related to articles 1 or 25 of the Brasilia Protocol (MRE, 1999e). This protocol, signed in Brasilia in July 1991, provides the mechanisms to solve disputes resulting from the interpretation, application, and default on the provisions of the Treaty of Asunción, of agreements concluded thereunder, of the decisions of the CMC, and of resolutions of the GMC. The Brasilia Protocol provides three procedures for extrajudicial dispute solutions: (1) negotiation (Chapter II), (2) conciliation (Chapter III), and (3) arbitration (Chapter IV).

The Joint Parliamentary Commission

The CPC represents the parliaments of the member countries. It is composed of a maximum of 64 acting parliamentary members—16 per member state and an equal number of alternates—appointed by the congress to which they pertain, and with a term of office of at least two years. The CPC meets ordinarily twice a year and additionally whenever asked by the president of any member country. The CPC's responsibilities are both advisory and decision-making in nature, and include assisting in the harmonization of laws of the different member countries and speeding internal national procedures to allow the prompt implementation of harmonized MERCOSUR standards.

The Socioeconomic Advisory Forum

This FCES represents the various socioeconomic sectors of the member countries and has a consultative nature.

The MERCOSUR Administrative Office

Located in Montevideo, Uruguay, SAM is administered by a national member, elected by the GMC on a two-year rotating basis. It keeps MERCOSUR documents and is responsible for the publication of all decisions adopted by the member countries and for the communication of activities of the GMC.

THE FOOD COMMISSION

Food law harmonization has been conducted by the SGT-3 (Technical Regulations Work Subgroup) under the responsibility of the Food Commission. To carry out its mission, the Food Commission is assisted by ad hoc groups whose activities have been prioritized (MRE, 1999f). These are given in Figure 6–2.

Originally, as stated in the Las Lenas Cronogram, the objective was to have all food law harmonized in time for the start of the single market (ABIA, 1997a). Not only were very few MERCOSUR resolutions adopted by January 1, 1995, but of those, very few were implemented in the member countries at that time.

In order to move toward economic integration, a common methodology for the elaboration and revision of MERCOSUR standards was needed. This is contained in MERCOSUR/GMC/Res. 152/96 (MRE, 1999g). It includes a provision establishing 180 days as the maximum period for the member countries to implement GMC technical regulations, which will then be mandatory. Applicable administrative measures are also provided. Before being submitted to the GMC, MERCOSUR technical regulations, which have been prepared and agreed to by the work subgroups, can be discussed by interested parties within each member state through an internal consultancy mechanism. This consultation must be concluded within 60 days of the ordinary meeting during which the project was approved by SGT-3.

In summary, the steps currently followed before the adoption of a MERCOSUR resolution by the member countries are:

1. elaboration of national proposals to be discussed jointly by governmental and private institutions

Southern Common Market Standards 85

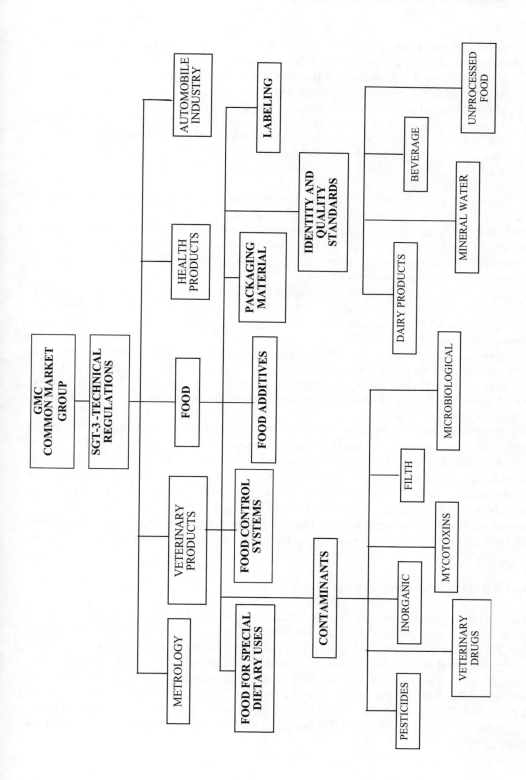

Figure 6-2 Composition of the MERCOSUR Food Commission

2. submission of the proposals to all member countries during an ordinary meeting of the SGT-3—Food Commission
3. discussion of the proposal by the specific ad hoc group
4. approval of the proposal by consensus
5. elaboration of a MERCOSUR project of technical regulation
6. internal discussion of the project within each member state by all interested parties (this step is not mandatory; the mechanism adopted by Brazil, for example, is the publication of the project of technical regulation in the official national press for public consultation during a 60-day period)
7. approval of the project by the SGT-3
8. submission of the harmonized project to the GMC for approval as a resolution

Once adopted, the GMC resolution must be introduced into the legislation of the individual countries within 180 days from the date of its publication by the GMC. The procedure for implementation varies according to the specific requirements of the legal system of each member state.

To supplement the scientific knowledge required to set food standards, Codex Alimentarius standards, guidelines, and recommendations as well as EU directives and U.S. Food and Drug Administration (FDA) regulations are consulted. Although the work of the Codex Committees and European Commission have progressed to adopt general horizontal measures to replace the detailed provisions contained in many of their commodities standards, there has been no consensus on this approach within MERCOSUR. In addition, delegations that may have agreed to change their positions at the ordinary meetings do not always make the same decision at the next meeting.

An analysis of the resolutions that have been adopted by the member countries shows that for some issues ad hoc groups have used their own criteria to judge which aspects should be included in the regulation. Also, in the early work of the Food Commission, standards were discussed that could have little direct influence on trade or health. With four countries involved, each with different laws, habits, and interests, harmonization has not progressed as rapidly as planned. Nevertheless, a number of regulations are now harmonized; these include some related to food additives and contaminants, packaging material, and dairy commodity standards.

Food Additive Ad Hoc Group

From the beginning of the harmonization process, the Food Additive Ad Hoc Group agreed that additives must be both technologically necessary and safe. MERCOSUR/GMC/Res. 31/92, the first recommendation of the member countries concerning food additives, defines food additives and establishes the fundamental principles for their use (MRE, 1999h). Based on national laws in force in the members states and on inclusion and exclusion criteria previously established and adopted as MERCOSUR/GMC/Res.17/93 (MRE, 1999i), a positive list of food additives was approved (MERCOSUR/GMC/Res.19/93 and subordinate resolutions) (MRE, 1999j).

Criteria for the approval of additives establish that provisions for food additives should take into account recommendations of the Codex Alimentarius and EU directives. FDA regulations are considered as supplementary references.

Functional classes of food additives and processing aids have also been defined (MERCOSUR/GMC/Res. 83/93 and Res. 84/93) (MRE, 1999k). The functional class titles adopted for food additives were acidulant, acidity regulator, anticaking agent, antifoaming agent, antioxidant, bulking agent, color, color stabilizer, emulsifier, firming agent, flavoring agent, flavor enhancer, flour treatment agent, foaming agent, gelling agent, glazing agent, humectant, preservative, raising agent, sequestrant, stabilizer, sweetener, and thickener. These classes of food additives are not completely harmonized with those adopted by Codex under the International Numbering System and do not necessarily reflect the national classification previously in force in the member countries. In Brazil, for instance, of the 23 classes of food additives adopted, 8 did not exist in the national legislation, and 4 were previously classified as processing aids. Unlike the classifications by Codex and the Joint Expert Committee on Food Additives and Contaminants (JECFA), packaging gas, propellant, and release agent were not included as functional classes of food additives, but as processing aids. Also, although enzymes were included in the harmonized list of food additives, it has been recognized that enzymes should be considered as ingredients or processing aids rather than additives since some of them are destroyed or lose their functions after processing.

Food colors and flavoring agents were the subject of MERCOSUR/GMC/Res.14/93 and Res. 46/93, respectively (MRE, 1999l). The Flavoring Framework Resolution outlines categories of flavorings, specifications, limits for undesirable substances, permitted additives, regional flavoring vegetable species, and certain labeling requirements.

Since their adoption, the original positive list of food additives and the positive list of food colors have been amended by consensus whenever new technological needs are presented by the member countries or new toxicological data become available. Potassium bromate, for example, although included in the original list, was later withdrawn (MERCOSUR/GMC/Res. 73/93) due to the JECFA's recommendation that its use as a flour-treatment agent was not appropriate (MRE, 1999m).

Functional classes were assigned to the approved additives, subsequent to the positive list of food additives, based on their technological uses both nationally and internationally (MRE, 1999n). Further progress included the approval of a list of food additives for use in foods in general, according to good manufacturing practices, unless otherwise specified in a particular commodity standard (MRE, 1999o). The additives that were included in this category are those for which "not specified" or "not limited" acceptable daily intake was established by JECFA.

The Codex Committee on Food Additives and Contaminants (CCFAC) in its 29th meeting approved a similar recommendation on this issue, which was adopted by CAC (CAC, 1999a). This demonstrates how closely MERCOSUR follows Codex decisions regarding food additives.

The food categorization system was based on the main categories of the Codex Food Categorization System (CFCS) with the introduction of as many categories/

subcategories as needed to accommodate national and regional foods. All foods were initially divided into 20 categories that were further divided into subcategories designed to include particular foods with specific needs for additives. Categories 1 to 11 match those of the CFCS. Categories 12 to 20 have different category titles than those of CFCS and are as follows: 12—soups and broths; 13—salts, spices, and sauces; 14—protein products; 15—food for special dietary uses; 16—beverages, including alcoholic ones; 17—coffee and tea, including mate leaves infusion; 18—snacks; 19—desserts; and 20—fortified foods. It should be emphasized that this system was internally agreed on by the Food Additive Ad Hoc Group for assigning food additives that were subject to changes to commodities.

Agreement on the technological need for food additives in particular foods has been difficult to achieve, and therefore, harmonization has progressed slowly. Additives have been assigned one by one into their respective food categories/subcategories only after long and tedious discussions. Such slow progress can be attributed to member countries' different raw materials, processing and packaging equipment, storage and distribution conditions, consumer dietary patterns, and consumer expectations. In order to resolve conflicting positions, member countries agreed on general criteria for assigning additives and their limits. In August 1997, during an extraordinary meeting, these criteria were amended to include other provisions related to the use of food additives, resulting in a document, *Criteria to Assign Functions of Additive, Permitted Additive and Its Limit to All Food Categories*, which was adopted by all member countries (MRE, 1999p). It was also agreed that all ad hoc groups on compositional standards must adopt these criteria when considering the technological need for a food additive and its maximum limit in a particular food category or subcategory.

The approved criteria to harmonize maximum limits cover different situations, such as when the maximum limit proposed by one member country is not accepted by another, or when there is no international or national reference for the use of the food additive or for the proposed maximum limit. In all situations, Codex standards, EU directives, and FDA regulations are considered, and the final decision must be adopted by consensus. Because of the number of commodities to be harmonized, member states have selected certain food categories to be discussed first, based on their importance in regional trade.

Provisions related to permitted uses of food additives in eight food categories have already been agreed to. They are 3—edible ices; 6—cereal and cereal-based products; 13—sauces, seasonings, and spices; 16(2)—nonalcoholic beverages; 8—meat and meat products; 7—bakery wares; 5—confectionery; and 19—desserts. Food categories presently being discussed are 9—fish and fishery products, and 12—soups and broths. The approved standards list the category and subcategories of foods and all food additives permitted to be added in each specific class function, followed by the maximum limit, expressed in g/100 g or g/100mL.

For some particular technological functions, the list of permitted additives might be considered too long. As already stated, many variables were taken into account in the elaboration of the standards, and the final document reflects the needs of all member countries. From a toxicological viewpoint, the possibility of using alternative additives may be an advantage since it will result in lower intakes of each individual additive.

Since Codex does not have provisions for the use of food additives in nonstandardized food categories, EU directives have served as the main international reference, especially for assigning maximum limits to food colors. Nevertheless, when the Codex General Standard for Food Additives (GSFA) is fully developed and adopted, MERCOSUR standards will certainly be revised and amended to comply with this standard.

Chemical Contaminants Ad Hoc Groups

General principles for the establishment of maximum limits of chemical contaminants in food were adopted in the framework of MERCOSUR (MRE, 1999q). Maximum limits must be assigned for mycotoxins, inorganic contaminants, pesticides, veterinary drugs, and hazardous migrants from packaging material. Decisions on this issue will take into account, among other things, the following information:

- maximum limits already established at regional and/or international levels
- representative regional data on the incidence of the contaminant, background information on the detected problem, analytical data, and potential health problems
- food of major importance for trade among the member countries
- data on human exposure at regional levels
- toxicological information on the contaminant
- standards, guidelines, and recommendations of the CAC, EU, FDA, and other internationally recognized bodies
- literature data available
- good manufacturing practices
- regional technology and availability

Comparison with Annex I of the Codex General Standard for Contaminants and Toxins in Foods (GSCTF), adopted at step 8 by the CAC at its 22nd session, shows that MERCOSUR principles on this issue have met the criteria recommended by Codex (FSIC, 1998). Nevertheless, the harmonization or assignment of maximum limits to veterinary drugs and pesticide residues, mycotoxins, and inorganic and microbiological contaminants has been the subject of much discussion in the respective ad hoc groups.

Very few maximum limits have been harmonized. Of the many reasons for this, two have been more frequently alleged by the member countries.

1. Agreement would represent a retrograde step for a country that already has in force a framework of law based on international recommendations.
2. The limits may be too low and, as a consequence, a recommendation may represent an unachievable target for those countries that have few resources to ensure adequate control.

Veterinary Drug Residues

National laws for these veterinary drug residues either do not exist or differ greatly from one member state to another (i.e., on what is permitted or prohibited), thus making the harmonization process even more difficult. Consequently, prior to the establishment of maximum residue levels (MRLs), the ad hoc group on these contaminants agreed upon criteria to define priorities for the control of veterinary drug residues in foods of animal origin as well as criteria for the validation of analytical methodology for these substances (MRE, 1999r). In a further step, MRLs, methodologies, and detection limits were established for 22 veterinary drugs, based on information from Codex and EU (MRE, 1999s). Since its adoption, this resolution has been revised and amended in face of new scientific information. Sampling procedures are presently being discussed.

Member countries have also highlighted the need for setting, jointly with the Commission on Veterinary Drugs, a priority list of compounds to be studied, and the necessity and importance of improving and implementing modern analytical techniques.

Mycotoxins

Harmonizing maximum limits (MLs) for these contaminants has been difficult. MERCOSUR member countries have different national regulations for mycotoxins. Only one regulation has so far been agreed upon (MRE, 1999t); it specifies MLs for aflatoxins in liquid milk, powdered milk, raw and processed peanuts, peanut butter, whole corn, and corn flour for human consumption. It also contains references to sampling plans and analytical methodology. In some situations, national and regional factors were taken into consideration, resulting in MERCOSUR tolerances that are higher than international standards. As compared to MRLs currently under discussion within the CCFAC, the maximum limit of 20 µ/kg agreed for total aflatoxins (B1 + B2 + G1 + G2) in unprocessed peanuts is higher than the Codex maximum level of 15 µ/kg for total aflatoxin in peanuts intended for further processing (CAC, 1999b), and the maximum limit of aflatoxin M1 in liquid milk is 10 times the proposed draft MLs of 0.05 µ/kg (CAC, 1999c). Although mycotoxins are considered a priority matter for harmonization, no meeting of this ad hoc group has occurred since 1995.

Inorganic Contaminants

A similar position has been developed on the harmonization of inorganic contaminants. MRLs for arsenic, copper, tin, lead, cadmium, and mercury were agreed upon in certain raw and processed commodities (MRE, 1999u) only after several meetings and long discussions. The Codex GSCTF would have facilitated the negotiations, but it is not yet fully developed. Thus decisions were based mostly on regulations in force in the member countries.

Of the 10 food categories/subcategories for which lead MRLs were set, fats and oils, citric fruit juices, and liquid milk are the only ones included in the Codex GSCTF proposed draft maximum level for lead. Lead MRLs adopted for different commodities, with the exception of fat and oils (MRL = 0.1 mg/kg), are higher than the proposed Codex MRLs for similar foods. At present the work of this ad hoc group has been discontinued.

Pesticides

Codex Alimentarius was the international reference that guided member countries toward harmonization of their national laws for pesticide residues. In order to facilitate the negotiation, member countries agreed to and internally adopted certain criteria.

Harmonization of MERCOSUR controls on pesticide residues started with developing a list of food commodities that would have priority in setting residue limits. Based on their importance for food trade, the following products were selected: beans, rice, mate leaves, tomatoes, strawberries, potatoes, apples, pears, garlic, sweet peppers, and soy beans. The current national legislation relating to each product was compared, and 1,125 pesticide residue limits were identified to be harmonized, of which 14 percent (158) were set during the transition period.

Since the official start of the Common Market on January 1, 1995, the ad hoc Group on Pesticides has had no meetings.

Filth and Microbiological Contaminants

One of the major concerns is microbiological safety. Although included among the priority items to be negotiated, the ad hoc group is currently inactive. Only one resolution has been agreed upon for these contaminants (MRE, 1999v), which specifies general principles for the establishment of criteria and microbiological standards for food based on Codex Alimentarius and the International Commission on Microbiological Specification for Food (ICMSF). This resolution also lists the food commodities that must be submitted to microbiological control.

There is some resistance by some countries to adopting horizontal standards for microbiological contaminants; the argument being that the origin of food-borne disease is not yet well known and that there is not enough information, both from national authorities and Codex Alimentarius, to allow further progress on this matter. Nevertheless, other member countries have emphasized the need to reactivate the Microbiology Ad Hoc Group in order to elaborate on horizontally harmonized technical regulations; doing so may eliminate trade barriers and contribute to the setting and implementation of vertical standards.

Packaging and Other Material in Contact with Food

Laws already exist in the EU regarding materials and articles in contact with food. FDA regulations also provide valuable information on this topic. At the time that negotiations started in 1992, very few national laws were in force in two of the member countries to control these contaminants, and these laws were both outdated and incomplete. Nevertheless, the legislation in Brazil and Argentina had been based on the same international reference (Italian legislation) which, in a certain way, has facilitated the negotiations.

During the harmonization process standards for metallic, cellulose-based, and returnable plastic packaging material as well as methodologies for the control of monomers and additives were elaborated upon. In Brazil, for instance, the legislation in force before MERCOSUR, which dated back to 1977, had provisions to control only eight monomers. MERCOSUR legislation has now provisions for the control of 33 monomers and 62 additives, which are specified in almost two dozen regulations.

Harmonization criteria for inclusion and exclusion of polymers, resins, additives for plastic material, and cellulose-based material, for example, are specified in MERCOSUR/GMC/Res. 56/92, Res. 95/94, and Res. 19/94 (MRE, 1999w). These regulations have been implemented in the member countries and most have already been amended in the face of new needs.

Identity and Quality of Food

Although the Codex Alimentarius has decreased its emphasis on the standardization of specific food commodities and increased emphasis on the work of committees in areas common to most foods, success in accomplishing this goal in MERCOSUR is still being debated. Member countries have difficulties agreeing on common standards for similar foods with particular national characteristics, and for products that are traded under the same name. Since national legislation is substituted after MERCOSUR resolutions are internally implemented, and delegations wish to protect traditional foods, they frequently fail to reach agreement in many negotiations. This is similar to what occurred in the EU, where few compositional food standards were produced.

Dairy Products

Because of the importance of trade in dairy products among MERCOSUR countries, this food category was selected as the first to harmonize for compositional standards. Since its inception, the work of the ad hoc Group on Dairy Products has progressed despite great differences among vertical standards of the member countries. More than 30 standards for dairy products, cheeses in particular, have already been agreed upon and implemented nationally. Difficulties encountered in reaching consensus include regional ones in the composition of similar products and in the technological need for certain food additives in specific products. This is associated mainly with differences in raw material and processing technology. In view of the speedy development of standards before full consideration of the issues involved, some of the standards may have actually gone beyond the aims of protecting the health of consumers and will be revised.

Beverages

Definitions and descriptions of raw material and elaboration processes of various nonfermented alcoholic beverages, as well as a classification of wines, were adopted by the member countries (MRE, 1999x). Although vertical standards for both alcoholic (spirituous) beverages and nonalcoholic beverages (fruit juices) have been discussed, no compositional standard has so far been adopted. Standards for beer are being harmonized.

Mineral Water

Despite the priority given to this product, conflicting opinions and interests regarding the compositional standards for water have been responsible for the inactivity of this ad hoc group.

Unprocessed Food

Identity and quality standards have been agreed on for vegetable commodities, which account for member countries' laws and technical regulations from the EU, Chile, and the U.S. Department of Agriculture. This ad hoc group is composed of specialized subgroups that deal with various food commodities such as fruits and vegetables.

MERCOSUR technical regulations for onions and tomatoes (MRE, 1999y) have already been adopted and implemented nationally. Some points of contention relate to container sizes. Problems arising from differences in climate and deficiencies in handling, which result in differences in quality standards, have also interfered with progress.

Labeling

Labeling is probably the area that causes the most difficulty in food trade. Written material and labeling must be in Spanish or Portuguese, depending on the country to which the product is exported. The information provided must inform and enable buyers so that they can distinguish clearly between seemingly similar products that in fact have different characteristics. In some member countries, basic national requirements for labeling are already in force although not always in the form required. National provisions for nutritional labeling practically did not exist when the harmonization process started.

Agreements include those on the labeling of prepackaged foods, nutritional labeling, and provisions for the declaration of ingredients and food additives (MRE, 1999z). A proposal of resolution on nutritional claims, based and elaborated on recent Codex Alimentarius and FDA documents, is presently being discussed by the member countries.

Food for Special Dietary Uses

Negotiations on this have not yet been initiated. Since Brazil has recently adopted national regulations on foods for special dietary uses (Brazil Government, 1998), those regulations will probably be considered working documents for discussion at the MERCOSUR level.

Food Control Systems

Because of the diversity and complexity of matters related to food quality control, the harmonization of national regulations has not progressed as expected. Member countries' widely divergent views on the use and control of food commodities are also a barrier to reaching agreement. Although proposed general principles on hygiene may not result in significant change for advanced food-producing member countries, enforcement authorities need to be informed of and agree to monitoring requirements.

In this age of effective international communication and increasing movement of food commodities, what is safe and acceptable in one country should be safe and acceptable in all countries, provided standards of hygiene are adequate and products are properly labeled. The problem lies in how to recognize and trust that a control system is adequate and equivalent. All delegations claim that their countries have adequate and effective food control systems, but in reality, their actions do not justify these claims.

Although eating habits and traditions vary throughout the four member countries, food hygiene and food safety have become an important MERCOSUR issue, and every member state should have confidence in the quality and safety of each others' products.

Since no agreement has been reached on how to recognize equivalence, *General Conditions to Establish Equivalence of Food Control Systems* were adopted (MRE, 1999zz). It contains objectives; definitions of common terms, such as *equivalence, control system, transparency, inspection, certification,* and *risk;* as well as precautionary measures to ensure safe and wholesome foods. This resolution also includes general principles and criteria to establish equivalence, which reference Codex Alimentarius principles, standards, and mechanisms for the enforcement of the rules by the appropriate authorities in the member countries.

Discussions have taken such a long time and proposals have changed so frequently that it is difficult to determine the current intent. Nevertheless, it would appear that as participants in the Uruguay General Agreement on Tariffs and Trade negotiations, member countries endorsed improved international trading rules concerning health-related regulations and will be obliged to base their decisions on sound science.

Since imported and locally manufactured products intended for consumption in the member state and for movement in the single market must be controlled, laboratories must be made available for the development of analytical methodology. While developed countries have applied financial resources to strengthen systems for the control of imported commodities, many developing countries have put more resources toward export control systems.

FAO TECHNICAL ASSISTANCE

The need to initiate implementation of the WTO's Application of Sanitary and Phytosanitary Measures and Agreement on Technical Barriers to Trade has resulted in requests of the Common Market Group for Food and Agriculture Organization's (FAO's) technical assistance in the field of food safety and quality. The assistance that FAO has provided in the framework of MERCOSUR includes activities related to the establishment and improvement of national inspection and certification systems, the harmonization of national and regional regulations with internationally recognized food standards and codes of practices, and the organization/strengthening of Codex national committees. A new project on the evaluation of conformity to facilitate mutual recognition of food inspection and certification among MERCOSUR countries, which includes an important training component, has also been requested by the GMC and will be implemented under the framework of the FAO's Technical Cooperation Programme (CAC, 1997).

These and other activities that have been supported by FAO will certainly speed up and facilitate the decision-making process on many conflicting issues and help in the implementation of national food control programs in response to MERCOSUR challenges.

CONCLUSION

With the creation of the WTO, it has been recognized that any national food legislation should be based on international consensus. MERCOSUR member countries elected Codex Alimentarius as the international reference organization. Decisions should be based on up-to-date scientific information, be transparent, and represent a balance resulting from the combination of competing interests of the consumer, industry, and trade.

MERCOSUR laws and regulations have taken into account regional laws and international laws, including requirements of the international trade agreement. The goal is a single set of achievable and enforceable food regulations, applicable in all member countries. There has been focus on the common goals stated by the Treaty of Asunción. Member countries have worked to reach agreement on legislation that will protect public health and the environment, and enhance public confidence in the safety of the food supply.

The issues addressed in MERCOSUR legislation are complex. Many times it has been difficult to achieve consensus. Some issues have never been faced or regulated within the member countries and some of the laws presently in force at a national level do not adequately reflect scientific advances.

At present, the harmonization process is working, and there is a growing number of decisions to be implemented at the national level. Although it is too early to assess the successes and failures of MERCOSUR, it should not be forgotten that, analogous to harmonization in the EU, the Treaty of Asunción was extremely ambitious and by no means easily implemented. The aim was to integrate and open economies of four Latin American countries and create a regional block to compete with already established international blocks. Although language is not a significant problem among MERCOSUR member countries, countries reflect their historical development through different legislation, habits, traditions, and national sovereignty that, together with a spirit of competition between member countries, has interfered with decision making.

Despite this backdrop, member countries with outdated legislation have now the unique opportunity to develop and update food laws that will allow most MERCOSUR commodities to be traded without barriers, not only within the regional market, but also in the important international markets. In this sense, Codex standards, EU directives, and FDA regulations have played an essential role in achieving consensus.

Significant agreements designed to meet the needs of changing consumer demands have been made, and the acceptance of Codex standards by MERCOSUR member countries is testimony to the vital role of this international agency in promoting fair trade and a safe food supply.

REFERENCES

Asociación Latinoamericana de Integración. (1999). *Que és ALADI?* http://www.aladi.org/aladi.htm. Accessed November 26, 1999.

Associação Brasileira das Indústrias de Alimentos. (1997a). Cronograma de Las Leñas. In Compêndio MERCOSUL-Legislação de Alimentos e Bebidas (Brazilian Association of Food Industries—Compendium on Mercosur-Legislation on Foods and Beverages). São Paulo: Associação Brasileira das Indústrias de Alimentos.

Brazil Government. (1998). Portaria SVS/MS nos. 29, de 13 Janeiro de 1998. *Regulamento Técnico para Identidade e Qualidade de Alimentos para fins Especiais*; Portaria SVS/MS nos. 30, de 13 Janeiro de 1998, *Regulamento Técnico para Fixação de Identidade e Qualidade de Alimentos para Controle de Peso*; Portaria SVS/MS nos. 34, de 13 Janeiro de 1998, *Regulamento Técnico para Fixação de Identidade e Qualidade de Alimentos de Transição para Lactentes e Crianças de Primeira Infâncias*. All in Diário Oficial da União. Brasília: Secretaria de Vigilância Sanitária.

Codex Alimentarius Commission. (1997). *Report on Activities Related to Economic Integration and Harmonization of Food Legislation in the Region (including Harmonization with Codex Standards)*. CX/LAC 97/6. Rome: Codex Alimentarius Commission.

Codex Alimentarius Commission. (1999a). Additives Permitted for Use in Food in General unless otherwise specified, in accordance with Good Manufactoring Practice—Annex to Table 3. In *Joint FAO/WHO Food Standards Programme Report of the Twenty-Third Session of the Codex Alimentarius Commission*. ALINORM 99/37, Appendix VII. Rome: Codex Alimentarius Commission.

Codex Alimentarius Commission. (1999b). Draft maximum level and sampling plans for total aflatoxin in peanuts for further processing. ALINORM 99/37. In *Joint FAO/WHO Food Standards Programme. Report of the Twenty-Third Session of the Codex Alimentarius Commission*. Rome: Codex Alimentarius Commission.

Codex Alimentarius Commission. (1999c). Draft maximum level for total aflatoxin in milk. ALINORM 99/37. In *Joint FAO/WHO Food Standards Programme. Report of the Twenty-Third Session of the Codex Alimentarius Commission*. Rome: Codex Alimentarius Commission.

Food Safety Information Center. (1998). Annexes I, II and III to the General Standard for Contaminants and Toxins in Foods. In *General Standard for Contaminants and Toxins in Food*. Edited by D.G. Kloet. Waggeningen: State Institute for Quality Control of Agricultural Products (RIKILT-DLO).

Ministério das Relações Exteriores. (1999a). Comércio Intra-Mercosul. Exportações 1991–1998. MRE/SGIE/GETEC. *Estatisticas y Comercio* 13. http://www.mre.gov.br/getec/WEBGETEC/BILA/22/apendice.html. Accessed November 20, 1999.

Ministério das Relações Exteriores. (1999b). Divisão do Mercado Comum. O Mercosul Hoje: *Produto Interno Bruto*. http://www.mre.gov.br/mercosul/PIB97.htm. Accessed November 26, 1999.

Ministério das Relações Exteriores. (1999c). Brazil Trade Net. *Parte XXIII 23. Tratados Inernacionais.* http://www.dpr.mre.gov.br/DB023-po.htm. Accessed November 26, 1999.

Ministério das Relações Exteriores. (1999d). Divisáo do Mercado Comum. O Mercosul Hoje. http://www.mre.gov.br/mercosul.htm. Accessed November 26, 1999.

Ministério das Relações Exteriores. (1999e). Protocolo de Brasília para a solução de controvérsias MERCOSUL/CMC/DECOI/1991. http://www.mre.gov.br/getec/WEBGETEC/BILA/ESP/document/2-docu.htm. Accessed November 21, 1999.

Ministério das Relações Exteriores. (1999f). Pautas negociadoras de los subgrupos de trabajo, reuniones especializadas y grupos ad-hoc. MERCOSUR/GMC/RES no. 38/95. http://www.mre.gov.br/getec/WEBGETEC/bdmsul/zlc&ua/2GMC/2RESOLUC/1995/RES3895.htm. Accessed November 26, 1999.

Ministério das Relações Exteriores. (1999g). Diretrizes para elaboração e revisão de regulamentos técnicos Mercosul. MERCOSUR/GMC/RES no. 152/96. http://www.mre.gov.br/getec/WEBGETEC/bdmsul/zlc&ua/2GMC/2RESOLUC/1996/RES15296.htm. Accessed November 26, 1999.

Ministério das Relações Exteriores. (1999h). Definições de ingredientes, aditivo alimentício, coadjuvantes de elaboração, contaminante e os princípios fundamentais referentes ao emprego de aditivos alimentícios. MERCOSUR/GMC/RES no. 31/92. http://www.mre.gov.br/getec/WEBGETEC/bdmsul/zlc&ua/2GMC/2RESOLUC/1992/RES3192.htm. Accessed November 21, 1999.

Ministério das Relações Exteriores. (1999i). Critérios para manutenção da lista de aditivos. MERCOSUR/GMC/RES no. 17/93. http://www.mre.gov.br/getec/WEBGETEC/bdmsul/zlc&ua/2GMC/2RESOLUC/1993/RES1793.htm. Accessed November 21, 1999.

Ministério das Relações Exteriores. (1999j). Lista geral harmonizada de aditivos Mercosul. MERCOSUR/GMC/RES no. 19/93. http://www.mre.gov.br/getec/WEBGETEC/bdmsul/zlc&ua/2GMC/2RESOLUC/1993/RES1993.htm. Accessed November 21, 1999.

Ministério das Relações Exteriores. (1999k). Definições de funções de aditivos alimentícios. MERCOSUR/GMC/RES no. 83/93. http://www.mre.gov.br/getec/WEBGETEC/bdmsul/zlc&ua/2GMC/2RESOLUC/1993/RES8393.htm; Definições de funções de coadjuvantes de tecnologia. MERCOSUR/GMC/RES no. 84/93. http://www.mre.gov.br/getec/WEBGETEC/bdmsul/zlc&ua/2GMC/2RESOLUC/1993/RES8493.htm. Both accessed November 21, 1999.

Ministério das Relações Exteriores. (1999l). Lista geral harmonizada de corantes. MERCOSUR/GMC/RES no. 14/93. http://www.mre.gov.br/getec/WEBGETEC/bdmsul/zlc&ua/2GMC/2RESOLUC/1993/RES4693.htm; Regulamento Técnico Mercosul de aditivos aromatizantes/saborizantes. MERCOSUR/GMC/RES no. 46/93. http://www.mre.gov.br/getec/WEBGETEC/bdmsul/zlc&ua/2GMC/2RESOLUC/1993/RES4693.htm. Both accessed November 21, 1999.

Ministério das Relações Exteriores. (1999m). Retira el Bromato de Potasio de la Lista General Armonizada de Aditivos MERCOSUR. MERCOSUR/GMC/RES no. 73/93. http://www.mre.gov.br/getec/WEBGETEC/bdmsul/zlc&ua/2GMC/2RESOLUC/1993/RES7393.htm. Accessed November 21, 1999.

Ministério das Relações Exteriores. (1999n). Lista de aditivos alimentares com suas classes funcionais. MERCOSUR/GMC/RES no. 101/94. http://www.mre.gov.br/getec/WEBGETEC/bdmsul/zlc&ua/2GMC/2RESOLUC/1994/RES10194.htm. Accessed November 21, 1999.

Ministério das Relações Exteriores. (1999o). Regulamento técnico sobre aditivos alimentares a serem empregados segundo as boas práticas de fabricação (BPF). MERCOSUR/GMC/RES no. 86/96. http://www.mre.gov.br/getec/WEBGETEC/bdmsul/zlc&ua/2GMC/2RESOLUC/1996/RES08696.htm. Accessed November 21, 1999.

Ministério das Relações Exteriores. (1999p). Regulamento técnico sobre critérios para determinar funções de aditivos, aditivos e seus limites máximos para todas as categorias de alimentos. MERCOSUR/GMC/RES no. 52/98. http://www.mre.gov.br/getec/WEBGETEC/bdmsul/zlc&ua/2GMC/2RESOLUC/1998/RES05298.htm. Accessed November 21, 1999.

Ministério das Relações Exteriores. (1999q). Princípios gerais para o estabelecimento de níveis máximos de contaminantes químicos em alimentos. MERCOSUR/GMC/RES no. 103/94. http://www.mre.gov.br/getec/WEBGETEC/bdmsul/zlc&ua/2GMC/2RESOLUC/1994/RES10394.htm. Accessed November 21, 1999.

Ministério das Relações Exteriores. (1999r). Critérios para definir prioridades de controle de resíduos de princípios ativos de medicamentos veterinários em produtos de origem animal. MERCOSUR/GMC/RES no. 53/94. http://www.mre.gov.br/getec/WEBGETEC/bdmsul/zlc&ua/2GMC/2RESOLUC/1994/RES5394.htm; Critérios para a validação de metodologias analiticas para a determinação de princípios ativos de medicamentos veterinários (RMV) em produtos de origem animal. MERCOSUR/GMC/RES no. 57/94. http://www.mre.gov.br/getec/WEBGETEC/bdmsul/zlc&ua/2GMC/2RESOLUC/1994/RES5794.htm Both accessed November 21, 1999.

Ministério das Relações Exteriores. (1999s). Resíduos de princípios ativos de medicamentos veterinários em produtos de origem animal. MERCOSUR/GMC/RES no. 75/94. http://www.mre.gov.br/getec/WEBGETEC/bdmsul/zlc&ua/2GMC/2RESOLUC/1994/RES7594.htm. Accessed November 21, 1999.

Ministério das Relações Exteriores. (1999t). Regulamento Técnico Mercosul sobre limites máximos de aflatoxinas admissiveis no leite, amendoim e miho. MERCOSUR/GMC/RES no. 56/94. http://www.mre.gov.br/getec/WEBGETEC/bdmsul/zlc&ua/2GMC/2RESOLUC/1994/RES5694.htm. Accessed November 21, 1999.

Ministério das Relações Exteriores. (1999u). Limites máximos de tolerância para os contaminates inorgânicos. MERCOSUR/GMC/RES no. 102/94. http://www.mre.gov.br/getec/WEBGETEC/bdmsul/zlc&ua/2GMC/2RESOLUC/1994/RES10294.htm. Accessed November 21, 1999.

Ministério das Relações Exteriores. (1999v). Princípios gerais para o estabelecimento de critérios e padrões microbiológicos para alimentos. MERCOSUR/GMC/RES no. 59/93. http://www.mre.gov.br/getec/WEBGETEC/bdmsul/zlc&ua/2GMC/2RESOLUC/1993/RES5993.htm. Accessed November 21, 1999.

Ministério das Relações Exteriores. (1999w). Disposiciones generales para envases y equipamientos plásticos en contacto con alimentos. MERCOSUR/GMC/RES no. 56/92. http://www.mre.gov.br/getec/WEBGETEC/bdmsul/zlc&ua/2GMC/2RESOLUC/1992/RES5692.htm; Lista positiva de aditivos para vasilhames e equipamentos plásticos que estejam destinados a entrar em contato com alimentos. MERCOSUR/GMC/RES no. 95/94. http://www.mre.gov.br/getec/WEBGETEC/bdmsul/zlc&ua/2GMC/2RESOLUC/1994/RES9594.htm; Envases y equipamientos celulósicos en contacto com alimentos. MERCOSUR/GMC/RES no. 19/94. http://www.mre.gov.br/getec/WEBGETEC/bdmsul/zlc&ua/2GMC/2RESOLUC/1994/RES1994.htm. All accessed November 21, 1999.

Ministério das Relações Exteriores. (1999x). Definiciones relativas a bebidas alcoholicas. MERCOSUR/GMC/RES no. 20/94. http://www.mre.gov.br/getec/WEBGETEC/bdmsul/zlc&ua/2GMC/2RESOLUC/1994/RES2094.htm; Definições relativas às bebidas alcoólicas. MERCOSUR/GMC/RES no. 77/94. http://www.mre.gov.br/getec/WEBGETEC/bdmsul/zlc&ua/2GMC/2RESOLUC/1994/RES7794.htm; Destilado alcoólico simples. MERCOSUR/GMC/RES no. 143/96. http://www.mre.gov.br/getec/WEBGETEC/bdmsul/zlc&ua/2GMC/2RESOLUC/1996/RES14396.htm; Reglamento vitivinícola del MERCOSUR. MERCOSUR/GMC/RES no. 45/

96. http://www.mre.gov.br/getec/WEBGETEC/bdmsul/zlc&ua/2GMC/2RESOLUC/1996/RES04596.htm. All accessed November 27, 1999.

Ministério das Relações Exteriores. (1999y). Identidade e qualidade de tomate. MERCOSUR/GMC/RES no. 99/94. http://www.mre.gov.br/getec/WEBGETEC/bdmsul/zlc&ua/2GMC/2RESOLUC/1994/RES9994.htm; Identidade e qualidade de cebola: PIQ. MERCOSUR/GMC/RES no. 100/94. http://www.mre.gov.br/getec/WEBGETEC/bdmsul/zlc&ua/2GMC/2RESOLUC/1994/RES10094.htm. Both accessed November 27, 1999.

Ministério das Relações Exteriores. (1999z). Regulamento Técnico MERCOSUR para rotulación de alimentos envasados. MERCOSUR/GMC/RES no. 36/93. http://www.mre.gov.br/getec/WEBGETEC/bdmsul/zlc&ua/2GMC/2RESOLUC/1993/RES3693.htm; Declaración de ingredientes en la rotulación de alimentos envasados. MERCOSUR/GMC/RES no. 6/94. http://www.mre.gov.br/getec/WEBGETEC/bdmsul/zlc&ua/2GMC/2RESOLUC/1994/RES0694.htm; Rotulado nutricional de alimentos envasados. MERCOSUR/GMC/RES no. 18/94. http://www.mre.gov.br/getec/WEBGETEC/bdmsul/zlc&ua/2GMC/2RESOLUC/1994/RES1894.htm; Declaración de los aiditivos en la lista de ingredientes. MERCOSUR/GMC/RES no. 21/94. http://www.mre.gov.br/getec/WEBGETEC/bdmsul/zlc&ua/2GMC/2RESOLUC/1994/RES2194.htm. All accessed November 27, 1999.

Ministério das Relações Exteriores. (1999zz). Condições gerais para estabelecer a equivalência de sistemas de alimentos. MERCOSUR/GMC/RES no. 59/99. http://www.mre.gov.br/getec/WEBGETEC/bdmsul/zlc&ua/2GMC/2RESOLUC/1999/RES05999.htm.

SICE—Foreign Trade Information System. (1999a). Southern Common Market (MERCOSUR) Agreement (Also known as the Treaty of Asunción). Organization of American States—Trade Unit. http://www.sice.oas.org/trade/mrcsr/mrcstoc.stm. Accessed November 20, 1999.

SICE—Foreign Trade Information System. (1999b). *Southern Common Market. Protocol of Ouro Preto*. Organization of American States—Trade Unit. http://www.sice.oas.org/trade/mrcsr/ourop/index.stm. Accessed November 20, 1999.

Part III

Scientific and Other Legitimate Factors

CHAPTER 7

Exposure Assessment Supporting International Developments

Naomi Rees

EXPOSURE ASSESSMENT AS PART OF RISK ASSESSMENT

The Food and Agriculture Organization/World Health Organization (FAO/WHO) Expert Consultation on the Application of Risk Analysis to Food Standards Issues (FAO/WHO, 1995a) defined risk assessment as containing the four components of: (1) hazard identification, (2) hazard characterization, (3) exposure assessment, and (4) risk characterization.[1] Subsequent FAO/WHO consultations have also defined elements of risk management and risk communication. Some of these definitions are under review within several committees of the Codex Alimentarius Commission (CAC) while the majority have been adopted by CAC. Figure 7-1 shows the relationship between the components of risk analysis in greater detail. This chapter will focus on exposure assessments as an important part of risk assessment.

Exposure assessment has been defined (FAO/WHO, 1995a) as, "The qualitative and/or quantitative evaluation of the likely intake of biological, chemical, and physical agents via food as well as exposures from other sources if relevant." This definition highlights the requirement to evaluate the likely intake from all relevant sources. Many of the models developed at the international level fall short of being able to assess exposure fully, according to this definition, because they do not cover all possible sources or pathways or are based on unlikely scenarios. This type of

---

[1]*Hazard identification:* The identification of biological, chemical, and physical agents capable of causing adverse health effects that may be present in a particular food or groups of foods; *hazard characterization:* The qualitative and/or quantitative evaluation of the nature of the adverse health effects associated with biological, chemical, and physical agents that may be present in food. For chemical agents, a dose-response assessment should be performed. For biological or physical agents, a dose-response assessment should be performed if the data are obtainable; *exposure assessment:* The qualitative and/or quantitative evaluation of the likely exposure of biological, chemical, and physical agents; *risk characterization:* The qualitative and/or quantitative estimation, including attendant uncertainties, of the probability of occurrence and severity of known or potential adverse health effects in a given population based on hazard identification, hazard characterization, and exposure assessment.

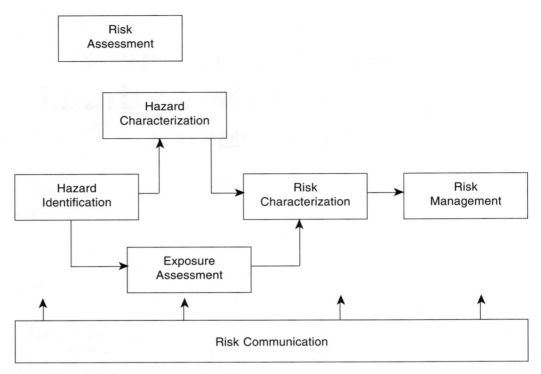

Figure 7–1 Risk Analysis Scheme as Developed by FAO and WHO (FAO/WHO, 1995a). *Source:* Data from *Application of Risk Analysis to Food Standards, Issues,* Report of the Joint FAO/WHO Expert Consultation, WHO/FNU/FO/95.3, © 1995, World Health Organization.

information is usually only available from national authorities. This definition also suggests that more than one approach can be used to estimate the dietary exposure of a chemical. The choice of approach will depend on the nature of data that are available (qualitative and/or quantitative) and the way in which a consumer is exposed to the chemical. Experience in the United Kingdom has shown that the method selected to estimate consumer exposure also depends on factors such as the quality of information available, the urgency with which the assessment is required, and the expertise available to carry out the assessment. In other words, a flexible approach to exposure assessment is the most effective one.

A range of methods for assessing dietary exposure is required. Estimates of dietary exposure can be based on different assumptions depending on whether long-term (chronic) or short-term (acute) effects of a chemical are being considered. International review of acute exposure assessment is at an early stage, but it is generally agreed that risk assessors need to have access to the most appropriate estimates of dietary exposure as well as toxicological advice, so that they can distinguish clearly between exposure patterns that could cause concerns on a long-term basis, and those that may indicate short-term problems. This distinction is also important because the way a food chemical risk is managed will differ depending on whether the chemical has chronic or acute effects. It will also depend on the severity and reversibility of the adverse effect, the probability of the hazard occurring, and the size and nature of the population group of concern.

Current approaches to exposure assessment at the international level cannot provide a full account of which consumers may be at risk of exceeding safety limits. This is because the dietary information available at the international level does not provide information on differences between individual consumers or the effect that, for example, age, sex, social class, season, or region can make to food consumption or exposure to a particular chemical. This may, however, be possible with national dietary survey information.

In the internationally agreed model of risk analysis (Figure 7–1), exposure assessment is part of risk assessment. This then provides the starting point for the risk management process. Characterizing risk is not easy, even with the best of information, and exactly how information is compiled and used by the risk manager is still under development in the management of food chemical safety. It is already clear, however, that the risk assessor needs to provide as full a picture as possible to the risk manager.

Estimation of exposure to food chemicals is a complex activity, and it is generally recognized that no single approach is suited to all circumstances. Information on dietary patterns and on the levels of chemicals in food is difficult and expensive to obtain. Approaches used to estimate consumer exposure to chemicals are reviewed in detail in Chapter 8, Methods for Estimating Dietary Exposure and Quantifying Variability.

NATIONAL DIETARY EXPOSURE ASSESSMENTS

In many countries the development of exposure assessment methodology was in progress long before the establishment of the World Trade Organization (WTO) in 1995. The Application of Sanitary and Phytosanitary Measures (SPS Agreement) and Agreement on Technical Barriers to Trade (TBT Agreement) contain statements about the importance of sound science supporting standard-setting activities. The influence of the SPS and TBT Agreements on food safety and international trade is reviewed in detail in Chapter 2, Food Safety and International Trade: The Role of the WTO and the SPS Agreement.

National exposure assessments have been requested by the Joint FAO/WHO Expert Committee on Food Additives (JECFA) (FAO/WHO, 1998a). There is no doubt that countries able to collate and provide detailed exposure assessment information have been able to participate to a greater extent than those that can not. Developing countries have not been able to contribute as much as they would like, partly because the whole process of assessing human health risk is so costly and partly because the infrastructure does not exist. Constraints experienced by developing countries are reviewed in detail in Chapter 9, The Need for Developing Countries To Improve National Infrastructure To Contribute to International Standards.

Governments in several countries are in the process of reorganizing their food safety policies in order to address current food safety issues more effectively. A greater emphasis is being placed on the use of resources and personnel to aid greater dissemination of information and more descriptive reporting procedures. The consumer's view of food safety has also changed. Chapter 10, A View from Consumers, describes the challenges.

From start to finish, a risk assessment can involve nutritionists, economists, mathematicians, chemists, toxicologists, regulators, managers, and industry personnel at different levels in a range of organizations. Risk assessment can also involve views from members of the public, input from different countries, reviews by several committees, comments by an established consultation procedure, and varying degrees of media attention. Therefore, it is important that there are clear lines of communication between the groups providing dietary survey information, those carrying out chemical surveillance activities, and others concerned with identifying, managing, and communicating food safety issues. It is also important that sufficient expertise, resources, and support are available throughout the process. This is not always the case and provides the greatest challenge.

INTERNATIONAL DIETARY EXPOSURE ASSESSMENTS

Before the mid-1980s, there were few, if any, formalized ways of assessing the implications for dietary exposure of standards being developed within international groups such as Codex. What international guidelines existed were difficult to put into practice because there was insufficient consumption information at the international level. Information collected through the Global Environmental Monitoring System (GEMS) on chemical exposure (WHO, 1985) was difficult to interpret because of the varying methods used in different countries. Standard-setting procedures were largely developed by consensus among the participating countries and based largely on levels requested by the relevant industry. National legislation, if it existed, influenced the outcome of the international standards, but was, in general, fairly weakly supported by detailed exposure assessments.

In the 1980s, because Codex standards were still being developed by committees responsible for a relatively small proportion of foods, estimates of exposure from the whole diet were not determined. In many countries, evaluation of the nutritional and health status of the population took greater priority than assessing the human exposure of nonnutrient chemicals in the diet. Total diet studies (also called market basket studies) were routinely used by several countries to assess the dietary exposure of nonnutrient chemicals such as pesticides and contaminants, and were published widely. However, the total diet studies were usually based on average diets. By the late 1970s and early 1980s, more consideration was being given to protecting at-risk (nonaverage) populations. Thus other exposure assessment methods were developed. During that time the whole area of risk analysis, as it applied to food safety, was also in its infancy.

At the international level, a number of discussions and publications in the late 1980s influenced thinking about exposure assessments and laid the groundwork for much faster development of this field in the early 1990s, which led to the FAO/WHO Consultation on Food Consumption and Exposure Assessment of Chemicals (exposure consultation), held in 1997 (FAO/WHO, 1998b).

Exhibit 7–1 summarizes the milestones in the development of dietary exposure methodology to support international standards. Guidelines for estimating the dietary exposure of contaminants, additives, and pesticides were published between 1985 and 1989 (CAC, 1989; WHO, 1985; WHO, 1989a). In 1990 a procedure for

Exhibit 7-1 Milestones in the Development of International Exposure Assessment Methodology Since 1985

Year	Outcome
1985	WHO published guidelines on dietary intake estimation of chemical contaminants. Three main methods were described: (1) the total diet study, (2) food diary methods, and (3) duplicate diet survey (WHO, 1985).
1988	WHO published guidelines on setting recommended levels for radionuclides in food, based on 11 hypothetical regional diets (WHO, 1988).
1989	CAC developed guidelines for evaluating additive intake. The first step considered per capita consumption and maximum permitted levels, and the next step considered per capita consumption and analytical data (CAC, 1989).
1989	GEMS/Food in collaboration with the Codex Committee on Pesticide Residues developed guidelines for estimating pesticide intake using per capita consumption values derived from FAO food balance sheet data. The first estimate considered the global diet and maximum residue levels, and then a regional diet with processing information, and ends by estimating intakes using national consumption and monitoring data (WHO, 1989a).
1990	JECFA developed a model diet of foods derived from animal tissue, eggs, and milk for use in setting MRLs for veterinary drug residues (WHO, 1989b).
1995	The FAO/WHO Expert Consultation on the Application of Risk Analysis to Food Standards Issues noted the importance of a harmonized approach to dietary exposure assessments across various Codex committees (FAO/WHO, 1995a).
1995	FAO and WHO convened a consultation, in May 1995 in York, United Kingdom (York Consultation), in an effort to review the 1989 pesticide intake guideline report (FAO/WHO, 1995b). The usefulness of the global diet was questioned and the diet was subsequently deleted.
1996	The first meeting of the JECFA Intake Group was held in February 1996 (WHO, 1997a).
1997	The FAO/WHO Consultation on Food Consumption and Exposure Assessment of Chemicals (Exposure Consultation) covered dietary exposure issues related to pesticide residues, food additives, veterinary drug residues, contaminants, and nutrients (FAO/WHO, 1998b).
1997	In June the CAC agreed that the five additives prioritized by the CCFAC should be considered by the JECFA Intake Group (CAC, 1997 and FAO/WHO, 1998a).
1997	WHO published revised guidelines for predicting dietary intakes of pesticide residues (WHO, 1997b).
1998	The CCFAC meeting in March 1998 considered an approach based on regional diets for estimating the dietary exposure of contaminants (CAC, 1998).

setting maximum residue levels (MRLs) for veterinary drugs was developed based on a model diet (WHO, 1989b). The pesticide guidelines prepared in 1989 were reviewed in 1995 (FAO/WHO, 1995b). There was considerable progress on the General Standard of Additives (Chapter 11, The Codex General Standard for Food Additives—A Work in Progress) and the General Standard for Contaminants and Toxins in Food (Chapter 12, Development of the Codex Standard for Contaminants and Toxins in Food).

Within the Codex standard-setting procedure, food is considered to be the primary source of exposure to chemicals. However, it is also important to assess total exposure by including drinking water and other routes of exposure such as air and occupational exposure (IPCS/WHO, 1998).

Exposure assessments have been, and still are, conducted differently from country to country. Although much of this is because each nation has tended to develop its own approach, there are also great differences in food consumption patterns between countries and, similarly, differences in the types and amounts of chemicals in food.

EXPOSURE CONSULTATION

The 1995 FAO/WHO Expert Consultation on the *Application of Risk Analysis to Food Standards Issues* noted the importance of a harmonized approach to dietary exposure assessments across Codex committees (FAO/WHO, 1995a). One of the objectives of the FAO/WHO Consultation on Food Consumption and Exposure Assessment of Chemicals (exposure consultation), held in 1997 (FAO/WHO, 1998b), was to promote consistent and transparent approaches for exposure assessment for the chemicals considered by Codex. The exposure consultation considered pesticide residues, veterinary drug residues, food additives, contaminants, and nutrients. The consultation agreed that, in principle, the determination of potential exposure is the same for all food chemicals although the specific application of procedures can differ. Table 7–1 provides an overview, in the Codex context, of the data currently used (i.e., food consumption and food chemical concentration), the methods by which exposure is estimated, the priority-setting functions, and issues related to overestimation, variability, and uncertainty.

The FAO/WHO exposure consultation (FAO/WHO, 1998b) defined international dietary exposure assessments as containing the following three main elements:

1. collation of data on food consumption (e.g., FAO food balance sheet data) or food chemicals (e.g., GEMS/Food [WHO, 1985] for residues of contaminants)
2. publication of guidelines that may range from designing dietary surveys to procedures for undertaking dietary exposure assessments for different food chemicals
3. development of dietary exposure methodology that can be used, on behalf of member countries of Codex, in order to screen and prioritize food chemicals for further work at the national level

Table 7–1 Overview of Dietary Exposure Assessments in Codex Alimentarius (FAO/WHO, 1998b)

	Pesticide Residues	Veterinary Drug Residues	Food Additives	Contaminants	Nutrients
Food Consumption Data	GEMS/Food regional diets (commodities)	Model diet of meat, kidney, liver, fat, eggs, and milk	Proposed* "unit quantity diet" approach (large portion weights of processed food)	Proposed** to use GEMS/Food regional diets (commodities)	Could use GEMS/Food regional diets (commodities)
	Codex Committee on Pesticide Residues (CCPR) food classification system	No classification system	Proposed food classification system based on the technological function of the food additive	Proposed to use CCPR food classification system (modified)	Dietary survey classification according to macronutrient content
	Large portion weights for acute assessment				
Food Chemical Concentration (or Residue Data)	MRLs	MRLs	Maximum levels (MLs)	MLs and Extraneous Residue Units (ERLs)	National food consumption databases and national food standards
	Supervised trials median residue (STMR) levels (chronic assessments) individual data (acute assessments)	GEMS/Food data	MLs submitted by industry (may take account of market distribution and substitution)	GEMS/Food data Could use median/mean contaminant level from survey data	Mean nutrient level (median level may be justified in some cases)
	GEMS/Food data				

*The "unit quantity diet" approach was discussed during the 29th CCFAC in March 17–21, 1997 (CAC, 1997).
**Under discussion at CCFAC.

continues

Table 7-1 continued

	Pesticide Residues	Veterinary Drug Residues	Food Additives	Contaminants	Nutrients
Method of Exposure Assessment	Theoretical maximum daily intake (TMDI) and IEDI	Model diet (back calculations to determine MRLs*)	Proposed screening: Budget Method followed by "unit quantity diet" approach; MLs; manufacturers' data	GEMS/Food regional diets with proposed MLs	None
	Compared with acceptable daily intake (ADI) or acute (RfD)	Compared with ADI	Compared with ADI	Compared with provisional tolerable daily intake/provisional tolerable weekly intake (PTDI/PTWI)	Compared with recommended daily intake (RDI) or health targets; PTDI/PTWI is usually the upper reference limit
	Additive effect(s) may need to be considered (e.g., organ-ophosphorus residues)	Other exposure routes may be considered (e.g., pesticide use, waste in environment)		Other exposure routes may need to be considered (e.g., heavy metals from air, work-related exposure)	Assessment of deficiency and toxicity differs (consider use as nutrient supplement or food additive)
Priority Setting Function	TMDI and, if data available, international estimated daily intake (IEDI) calculated	Only one method used	Proposed tiered approach to identify priorities	Proposed priority function for identifying commodities that require an ML for each contaminant	None (could be used internationally for prioritizing nutrients for labeling)

*Note that often the MRL is established using an indicator compound.

continues

Table 7–1 continued

	Pesticide Residues	Veterinary Drug Residues	Food Additives	Contaminants	Nutrients
Over-estimation, Variability, Uncertainty	GEMS/Food regional diets overestimate food consumption of commodities for the average consumer and may underestimate the consumption for high consumers	Model diet tends to overestimate exposure to veterinary drug residues	Budget Method and "unit quantity diet" approach may overestimate lifetime exposure	GEMS/Food regional diets overestimate food consumption of commodities for the average consumer and may underestimate the consumption for high consumers	Data too uncertain to undertake international estimates
	Uncertainties in processing factors	Food consumption and level of residue both overestimated	"Unit quantity diet" approach requires validation from countries	Uncertainties in processing factors	Variability in nutrient intake between countries depends on food composition and dietary patterns
	Variability in residue data depends on reporting system limits of determination (LOD), percent application, and percent coverage			Variability within and between countries; also depends on: reporting system; LOD	
				Require contaminant survey data to be reported consistently	Different groups at risk in different countries
	Require residue survey data; reported consistently				
	Uncertainties associated with derivation of ADI and the acute RfD	Uncertainties associated with derivation of ADI	Uncertainties associated with derivation of ADI	Uncertainties associated with derivation of PTDI/PTWI	Uncertainties associated with derivation of RDI and PTDI/PTWI

Source: Reprinted with permission from *Food Consumption and Exposure Assessment of Chemicals, Report of a FAO/WHO Consultation*, 10–14 February 1997, WHO/FSF/FOS 97.5, © 1998, World Health Organization.

FUNCTION OF INTERNATIONAL EXPOSURE ASSESSMENTS

The FAO/WHO exposure consultation agreed that, "At the international level, dietary exposure assessments have an important function to screen food chemical(s) intakes and identify those chemicals with a potential public health and safety concern."

The exposure consultation also noted that "while international dietary exposure assessments are useful for those countries with no capacity to undertake national dietary exposure assessments, such exposure assessments cannot be used to estimate actual dietary exposures to food chemicals in individual countries or for specific population subgroups."

The consultation acknowledged that increased accuracy in estimating dietary exposure carried with it an increased cost. Simpler methods can play a useful role, at the international level, *provided* they overestimate exposure (or they underestimate exposure if nutritional deficiency is the issue). However, these assessments should not pile conservative assumptions onto other conservative assumptions, thereby making international calculations ineffective as screening tools. It may be more important, at the early stages of the risk assessment, to demonstrate that the dietary exposure does not (or cannot) exceed the safety limit for chemicals. At the same time, an attempt should be made to provide the best estimate possible with the available information while discussing the potential uncertainties.

International estimates of dietary exposure should be based on realistic assumptions and sound science. As far as possible, the assessments need to consider regional variations in food consumption, differences in chemical levels found in food, and other factors such as nutritional or health status of the population. Where possible they must be checked against national estimates in order to assess the validity of the international methods and the areas of uncertainty.

By definition, any exposure assessment carried out at the international level cannot assess the situation in individual countries. International exposure assessments are not usually very useful when the issues are overly complex. They can take a considerable amount of time. The main function then of Codex member states would be to monitor the accuracy of these estimates and reconcile any differences. An international assessment is only as good as the underlying principles and the available information.

FACTORS INFLUENCING THE DEVELOPMENT OF INTERNATIONAL EXPOSURE ASSESSMENT METHODOLOGY

Considering the experience of the last 20 years or so, it would seem that the speed and effectiveness of dietary exposure assessment methods developed at an international level have been, and will be, influenced by the following factors:

- The availability of data. For example, regional diets (WHO, 1998) are used to estimate exposure to pesticide residues and have been proposed for veterinary drug residues and contaminants, but are not suitable for additives (FAO/WHO,

1998b). A budget method has been used as a first screen for additives (CAC, 1999).
- The collective resources of the Codex committees involved and their members. For example, the tremendous amount of resources required to support the development of the General Standard of Food Additives (GSFA) was provided by relatively few member countries of Codex.
- The ability to gain consensus on the underlying principles for carrying out and interpreting dietary exposure assessments. This is particularly difficult since Codex committees usually meet at most only once a year. Increasing the number of meetings would be costly and could alienate those members who could not attend.
- Ongoing discussions on the balance between trade and human health in risk analysis.
- Interactions between risk assessment committees (such as JECFA and JMPR) and risk management committees, such as Codex Committee on Food Additives and Contaminants and the Codex Committee on Pesticide Residues (CAC, 1999).

CONCLUSION

The FAO/WHO exposure consultation made a valuable contribution in discussing the role and function of international exposure assessment, by promoting a consistent framework and recognizing the limitations and areas for development. These concepts should mature as risk analysis principles are put into practice across the world.

REFERENCES

Codex Alimentarius Commission. (1989). *Guidelines for Simple Evaluation of Food Additive Intake*, 1st ed., Supplement to Codex Alimentarius Vol. XIV. Rome: Joint FAO/WHO Food Standards Programme.

Codex Alimentarius Commission. (1997). *Report of the 29th Session of the Codex Committee on Food Additives and Contaminants*. The Hague, the Netherlands, March 17–21, 1997, ALINORM 97/12A. Rome: Codex Alimentarius Commission.

Codex Alimentarius Commission. (1998). *Report of the 30th Session of the Codex Committee on Food Additives and Contaminants*. The Hague, the Netherlands, March 9–13, 1998, ALINORM 99/12. Rome: Codex Alimentarius Commission.

Codex Alimentarius Commission. (1999). *Report of the 31st Session of the Codex Committee on Food Additives and Contaminants*. The Hague, the Netherlands, March 22–26, 1999, ALINORM 99/12A. Rome: Codex Alimentarius Commission.

Food and Agriculture Organization/World Health Organization. (1995a). *Application of Risk Analysis to Food Standards Issues*. Report of the Joint FAO/WHO Expert Consultation. WHO/FNU/FOS/95.3. Geneva, Switzerland: World Health Organization.

Food and Agriculture Organization/World Health Organization. (1995b). *Recommendations for the Revision of Guidelines for Predicting Dietary Intakes of Pesticide Residues*. Report of a FAO/

WHO Consultation, May 2–6, 1995, York, United Kingdom. WHO/FNU/FOS/95.11. Geneva, Switzerland: World Health Organization.

Food and Agriculture Organization/World Health Organization. (1998a). Joint FAO/WHO Expert Committee on Food Additives. *List of Substances Scheduled for Evaluation and Request for Data.* Fifty-first meeting. Geneva, June 9–18, 1998. Annex 3. Geneva, Switzerland: World Health Organization.

Food and Agriculture Organization/World Health Organization. (1998b). *Food Consumption and Exposure Assessment of Chemicals.* Report of a FAO/WHO Consultation, February 10–14, 1997, Geneva. WHO/FSF/FOS/97.5. Geneva, Switzerland: World Health Organization.

International Program on Chemical Safety/World Health Organization. (1998). *International Program on Chemical Safety, Environmental Health Criteria. Human Exposure Assessment.* UNEP/ILO/WHO, PCS/EHC/98.4. Geneva, Switzerland: World Health Organization.

World Health Organization. (1985). *Guidelines for the Study of Dietary Intakes of Chemical Contaminants.* Offset Publication. No. 87. Geneva, Switzerland: World Health Organization.

World Health Organization. (1988). *Derived Intervention Levels for Radionuclides in Food.* Geneva, Switzerland: World Health Organization.

World Health Organization. (1989a). *Guidelines for Predicting Dietary Intake of Pesticide Residues.* GEMS. Geneva, Switzerland: World Health Organization.

World Health Organization. (1989b). *Evaluation of Certain Veterinary Drug Residues in Food, Thirty-Fourth Report of the Joint FAO/WHO Expert Committee on Food Additives.* Technical Report Series, No. 788. Geneva, Switzerland: World Health Organization.

World Health Organization. (1997a). *Forty-Sixth Report of the Joint FAO/WHO Expert Committee on Food Additives.* Technical Report Series, No. 868. Geneva, Switzerland: World Health Organization.

World Health Organization. (1997b). Guidelines for Predicting Dietary Intake of Pesticide Residues, revised. WHO/FSF/FOS/97.7. Geneva, Switzerland: World Health Organization.

World Health Organization. (1998). *GEMS/Food Regional Diets, Regional per Capita Consumption of Raw and Semi-Processed Agricultural Commodities.* WHO/FSF/FOS 98.3. Geneva, Switzerland: World Health Organization.

CHAPTER 8

Methods for Estimating Dietary Exposure and Quantifying Variability

Barbara J. Petersen

INTRODUCTION

The mere ingestion of a chemical does not imply any adverse health consequence. In fact, some chemicals are essential nutrients. However, virtually all chemicals are toxic at some dose. For example, many essential nutrients are toxic at levels higher than the levels that are essential. Therefore, the amount of the chemical that is ingested and its toxicity profile should guide the understanding of the significance of the intake. Although the interpretation of the results may be different, the methodologies for estimating intake are similar for toxins, nutrients, and microorganisms.

There is variability in the intake of any chemical. Also, the diets of individuals vary—both between individuals and in the same individual from day to day. This chapter focuses on methods for measuring the intake and for quantifying variability.

OVERVIEW OF DIETARY METHODS AND DATA AVAILABLE FROM DIFFERENT COUNTRIES

In order to assess intake, four types of data are required:

1. potential levels in each food
2. frequency of occurrence of the chemical
3. absorption factors
4. amounts of each food that are consumed by the population being evaluated

There are many sources of data and a variety of methods that can be used to determine the contribution of ingestion to the total intake (Petersen & Douglass,

1994; Petersen & Barraj, 1996; Trichopoulou & Lagiou, 1997). Regardless of the methods and data, all procedures estimate intake using the following equation:

$$\text{Intake} = \text{concentration in media} \times \text{consumption of media (quantity and frequency)} \times \text{proportion absorbed}$$

The selection of the most appropriate methodology will depend on: (1) the intended application for the intake assessment, (2) the biological properties, and (3) the physical and chemical properties of the substance. In February 1997, Food and Agriculture Organization/World Health Organization (FAO/WHO) (WHO, 1998) convened a joint consultation to develop procedures for estimating dietary intake at the international level. Two major outcomes of the conference were: (1) schemes for evaluating intake that proceed from screening techniques at the international level to refined intake analyses at the national level; and (2) revised procedure for developing regional diets for use in risk assessments for pesticides and contaminants. Some of the important considerations for each of the two areas were summarized in a graphic from that report (reproduced in Figure 8–1) and are discussed below.

The Intended Application of the Intake Assessment

The purpose of the assessment plays a critical role in determining the most desirable methodology. The optimum method when the assessment is designed to be conservative (as is often the case for regulatory decision-making applications) is different from the optimum method when the analysis is designed to be as realistic as possible (as in scientific hypothesis testing).

Some methods, such as the Budget Method (Hansen, 1979), have been designed as screening methods. This method yields a worst case estimate that is often used internationally to screen food additives. This method relies on estimates of physiological requirements for total food and liquid rather than on food consumption survey data. It does not provide an estimate of intake of the additive. Although the Budget Method dramatically overestimates intake, as a preliminary assessment it can be used to prioritize resources and to design sampling programs.

Screening methods and model diet methods sacrifice accuracy, in estimating intake, for speed and simplicity. In the case of the evaluation of toxic effects, screening results that are acceptable indicate that actual intakes will be acceptable since they will be lower than the worst case estimate. Therefore, it can be assumed that there is no need to expend resources to collect better data or to apply more sophisticated techniques in search of greater accuracy. In contrast, a research project that is attempting to evaluate the cause-and-effect relationship of a chemical and a disease would require more accurate intake assessments.

The Biological Properties of the Chemical

The length of dosing that is required to elicit a specified biological effect should be used to define the key parameters for intake assessment. That is, the biological effects that are the result of a single or at most few doses will be compared to dietary

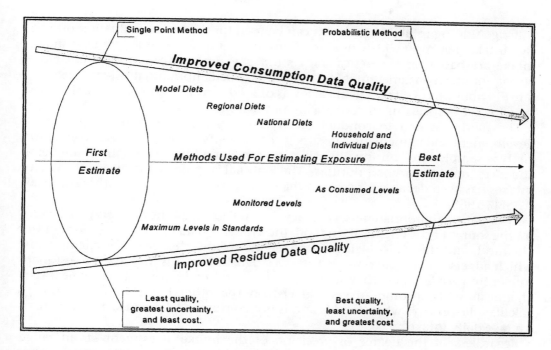

Figure 8–1 Relationship between Food Consumption, Food Chemical Concentration Data, and Assessment Method and the Quality, Uncertainty, and Cost of the Resulting Dietary Exposure Assessment. *Source:* Reprinted with permission from *Food Consumption and Exposure Assessment of Chemicals, Report of a FAO/WHO Consultation*, 10–14 February 1997, WHO/FSF/FOS 97.5, © 1998, World Health Organization.

intake on a single day. Correspondingly, toxic effects that arise as a result of long-term intake will be compared to average dietary intakes (usually over a year).

Other considerations include any breakdown products of toxicological significance, and the metabolic pathways in plant and animal systems. Potential biological effects must be carefully considered in planning an intake assessment. Factors of interest include dose-response relationships, the length of intake required to produce an adverse effect, sensitive populations, variability, and uncertainty factors.

The Physical and Chemical Properties of the Chemical

Often when estimating intake of a food additive or contaminant, it is necessary to define or characterize the chemical in terms of attributes such as structure, volatility, and solubility. For example, does the substance break down during storage, processing, and cooking?

Diets in many countries are highly processed. Therefore, for most assessments, it will be critical to include estimates of the residues in the products as they are consumed (Chin, 1991).

Range-finding intake assessments can be used for the initial evaluation of worst case intake. The most widely used methods to determine the order of magnitude of intake are based on theoretical maximum chemical concentrations and either average or high consumer intakes. These initial calculations usually do not incorporate distributions of either chemical levels or food consumption patterns; similarly, the effects of processing or cooking are not included.

The methods used to conduct dietary surveys are in a continuing state of development and refinement. The survey instruments and procedures to collect and analyze samples are continually being improved, and new compounds are being added to reflect current priorities. The optimal approach for assessing dietary intakes has been debated for years in the United States as well as in other countries (FASEB, 1988).

The relative appropriateness, accuracy, and reliability of the analysis will depend on the source of the chemicals in food, the available data for concentrations of the chemical, and the toxicological characteristics of the chemical. For chemicals in which adverse effects are observed after a single intake, it is important to estimate intake by people when they actually ingest food containing the chemical. For chemicals with chronic adverse effects, on the other hand, intake is usually calculated over days when food is and is not consumed to produce an estimate of average daily intake.

Regardless of the degree of precision of the intake assessment—from range finding to precise estimates—it is critical that the most appropriate intake assessment model is applied. The model that is selected will also help determine which food consumption survey is most appropriate.

There are three general intake assessment models.

1. **Average intake**—The most basic models combine data on average food consumption and average concentration levels of the chemical to estimate average intake. Average chronic intake is usually estimated on a per capita consumption basis and is compared to the measurements of biological/toxicological results from lifetime animal feeding studies or other appropriate test results. Single day or acute intakes may be computed using similar methods, but using food intake data for a single meal or for the day. Acute intakes are usually compared to results of tests in which the subjects were dosed for a single day.
2. **Simple distribution**—A simple distribution of intake can be calculated by applying a single, average (or worst case) estimate of the chemical's concentration level to a distribution of food consumption. Alternatively, a distribution of concentration levels can be applied to an average or worst case food consumption level.
3. **Joint distribution/Monte Carlo-type probabilistic assessment**—In creating a joint distribution to estimate intake, the distribution of food consumption is combined with the distribution of chemical concentration levels. Joint distribution/probabilistic analysis allows the most realistic estimates of intake.

DEFINING THE INTAKE SCENARIOS

The sources of intake must be defined along with the determination of magnitude and frequency of intake. It is usually desirable to know how the chemicals enter the food supply and the time frames that are involved. The following steps help to define the most appropriate intake scenario:

1. Identify the regulatory standards that apply to the chemical (e.g., WHO, European Union, or U.S. maximum limits).
2. Determine the characteristics of intended or accidental additions of the chemical to the food supply, to food contact surfaces, or onto the hands. Determine the impact of processing/cooking and storage of food on the final levels of the chemical.
3. Identify and evaluate existing data for both the levels of the chemical and for the appropriate consumption estimates.
4. Create or apply draft intake assessment algorithms that define the parameters to be computed and the models to be used to estimate intake. If appropriate, develop a tiered methodology that allows a sequence of analyses from range finding to detailed probabilistic intake assessments.
5. Conduct appropriate preliminary intake assessments including sensitivity analyses to assist in defining the parameters, which need the best available data and/or additional data generation.

Quantifying the Concentration of the Chemical for Food Groups or for Individual Foods

The analyst's goal is to characterize the concentration of the chemical in each food. To assess intake of a chemical, it is important to consider not only the average concentrations, but also the variability in concentration.

Existing data can be used to estimate intake and to guide the collection of future data. A substantial amount of information that is relevant to intake assessment is available for some chemicals (Graham et al., 1992 and Sexton et al., 1992). Such information can be obtained from existing databases on food and drinking water consumption and contaminant residues normally maintained for other purposes (i.e., nutritional intake or regulatory surveillance). The Total Diet Study (Pennington & Gunderson, 1987) of the Food and Drug Administration (FDA), the Nationwide Food Consumption Survey of the U.S. Department of Agriculture (USDA) (USDA, 1991 and 1996), the National Health and Nutrition Examination Surveys (NHANES) of the U.S. National Center for Health Statistics (NCHS, 1982), as well as other national regulatory monitoring programs, can provide information useful for dietary intake modeling and assessments.

Useful data sources for defining the foods with the highest potential concentration of the chemical include:

- chemical properties (solubility, heat stability, pH stability, and other properties)
- metabolites/degradates of potential concern
- differences in levels of raw versus cooked food
- levels in home-prepared versus commercially processed foods
- time frames for presence of chemical in the food supply or in categories of the food supply—continuously or in periodic intervals
- impact of processing/home preparation on potential residues

There are three common approaches that are appropriate for collecting data on concentrations of chemicals in food: (1) duplicate diets, (2) market basket or representative sampling, and (3) controlled experimentation. These are discussed below.

Duplicate Diets

Canada, United Kingdom, Japan, and Sweden have successfully used daily intake measurements with a high degree of success in both regulatory and assessment programs (e.g., Dabeka et al., 1987; Sherlock et al., 1983; and Vahter et al., 1990).

To make an accurate assessment of dietary consumption, the actual foods being eaten must be collected and analyzed, both in and away from the home and in commercial and occupational eating places. The collection of duplicate diets has proven to be a viable, if expensive, sampling approach. Duplicate diet procedures were the method of choice for the U.S. Environmental Protection Agency's National Human Exposure Assessment Survey (FASEB, 1993) and the WHO's Global Exposure Monitoring System (WHO, 1983).

In duplicate diet sampling, a duplicate portion of all foods and beverages consumed during the monitoring period is collected. Foods collected are normally composited for analysis, because of cost, to yield a single measure of dietary intake of the participant for a daily or longer monitoring period. Variations of these procedures are used depending on study objectives. Individual food items or food groups and daily food collections may be segregated for independent analysis. Alternatively, it may be necessary to sample foods at the various steps during food preparation in order to identify the reason for excess contamination. Food diaries and questionnaires are used to obtain supporting information useful for identifying sources of contamination and for measuring the completeness and representative nature of the foods collected. Duplicate diet collections should include all foods consumed, both in and away from the residence (FASEB, 1993), and the monitoring period should be extended to the maximum number of days possible without undue burden on the participant. The researcher must determine the level of information required for various intake scenarios. Often detailed intake information can only be obtained at considerable expense and participant burden. For example, detailed dietary histories, segregation of individual food items for analysis, and weighing individual foods may not be necessary when a single measure of total intake is the goal.

Market Basket and Surveillance Sampling

Many countries conduct various types of monitoring or surveillance for chemicals in commodities or in foods (WHO, 1997). Monitoring and surveillance studies are conducted to assess compliance with state, national, or international regulations.

At the U.S. federal level, the USDA monitors residue levels in selected fruits and vegetables, meat and poultry products. The FDA monitors residue levels in all other foods. California, Florida, and a number of other states have monitoring programs. Depending on the specific U.S. monitoring program, foods or commodities may be sampled at the point of entry to the country, at the farm gate, at the food processing plant, or at retail level.

Market basket surveys are conducted in the United States and other countries to obtain food chemical concentration data that may be used in intake assessment. A core group of foods that are representative of national dietary patterns is obtained and analyzed to determine the concentrations of the substances of interest. Generally, samples of food are purchased at retail outlets in different regions of the country and prepared as for consumption.

Experimental Results

Controlled experiments are sometimes used to determine the likely levels of chemicals in foods produced under specified conditions and of relevant metabolites. Controlled experiments have also been conducted to identify the source of other chemicals in foods. For example, studies have been conducted to quantify the migration of lead from pottery and glassware into foods and beverages (Bolger et al., 1996).

DIETARY INTAKE

In the United States, Europe, Australia, and many other countries (Australia New Zealand Food Authority [ANZFA], 1997; SCF, 1994) good food consumption survey data are available. However, these surveys are conducted for purposes other than intake estimations. Commodity monitoring data are used to track agricultural production and imports. Food consumption data on individuals or groups of individuals are collected to characterize expenditures, dietary intake, or the nutritional status of those individuals.

The most appropriate survey for a particular assessment must be determined. There are a number of specific questions to consider when selecting food consumption data to estimate intake:

- When (how long ago) were the data collected?
- Are current patterns of consumption of the foods of interest similar enough to patterns existing when these data were obtained?
- For which population were the food consumption data collected?
- For which country or countries were the food consumption data collected?
- For which geographical regions were the food consumption data collected?
- Were all population groups included (e.g., children, vegetarians, and subsistence fishermen)?
- Were data collected during all seasons?

- What foods were included?
- Was the quantity of each food estimated?

Information about the level of detail in food consumption data available from a particular survey is an important determinant of the potential for accuracy in estimating intake of a specific chemical.

Types of Food Consumption Surveys

There are four broad categories of food consumption data: (1) food supply surveys (market disappearance), (2) household or community inventories, (3) household food use, and (4) individual food intake surveys.

Food Supply Surveys

Food supply surveys, also called food balance sheets (FBSs) or disappearance data, describe a country's food supply during a specified time period. FBSs published by the FAO describe the food supply in countries on all continents. European FBSs are also prepared by the Organization for Economic Cooperation and Development (OECD) and the Statistical Office of the European Communities (EUROSTAT). Food supply data in the United States are developed by USDA's Economic Research Service.

Availability is determined at different points for different foods or commodities. Mean per capita availability of a food or commodity is calculated by dividing total availability of the food by the total population of the country.

These surveys provide data on food availability or disappearance rather than actual food consumption, but may be used to estimate indirectly the amounts of foods consumed by the country's population. Food supply data may be useful for setting priorities, analyzing trends, developing policy, and formulating food programs. For some countries, food supply data are the only accessible data representing the country's food consumption. Because similar methods are used around the world, these data may be used to make international comparisons and may also be useful in some epidemiological studies (Sasaki & Kestelhoot, 1992).

WHO Global Environment Monitoring System—Food Contamination Monitoring and Assessment Programme (GEMS/Food) developed a series of global and regional/cultural diets to use in conducting risk assessments at the international level (WHO, 1988; WHO 1998). These diets are based on FAO FBSs from selected representative countries as well as on expert interpretation. For more than 10 years, the diets provided the food consumption component of the dietary intake calculations that have been used by WHO for predicting dietary intake of pesticide residues for the Codex Committee on Pesticide Residues (CCPR) and the Joint Meeting on Pesticide Residues (JMPR).

The Joint FAO/WHO Consultation on Guidelines for Predicting Dietary Intake of Pesticide Residues (FAO/WHO, 1995) recommended that the WHO GEMS/Food global diet be discontinued and the five existing regional/cultural diets be revised. This recommendation was endorsed by the CCPR and the JMPR. The revision of the regional/cultural diets was further discussed at the Joint FAO/WHO Consultation on Food Consumption and Exposure Assessment of Chemicals (FAO/WHO, 1998). A new approach for developing the regional diets, including assigning countries to a specific region, was proposed, based on initial work using a statistical cluster method.

Following that guidance, the FAO FBSs have been subjected to cluster analysis to guide the updating of the WHO global diets used for pesticide intake assessments (Barraj et al., 1999). Data from FAOSTAT.PC, Version 3.0, 1996 were used. The data available for each country are not necessarily consistent across years because new countries were added to the database after 1992 (e.g., some of the newly independent states that were previously part of the USSR), while some were dropped (e.g., the USSR). The consumption data are on a per capita basis and are expressed in terms of quantity, caloric value, fat, and protein content. The data cover a period ranging from 1965 to 1994. The consumption data (in kg/year) for all *individual* countries from the 1990 to 1994 FAO balance sheets were used in the derivation. The quantity of each food was selected as the intake estimate.

The approach used to derive "cultural" diets consisted of a three-stage process. At the first stage, the statistical method known as cluster analysis (Romesburg, 1990) was used to divide the 182 countries into nine statistical clusters based on the similarity in their consumption levels of 19 marker foods and food groups (Figure 8–2). The 19 foods were selected in a two-stage process, whereby 6 food groups representing the major foods were first selected (i.e., cereals, fruit, vegetables, meats, offal, and fish). This list was then arbitrarily extended by adding some of the major foods included in the food groups. The k-means clustering method was used. The method first selected nine countries as cluster seeds. These are selected in order to have large differences between them with respect to their consumption levels. The remaining countries are then assigned, in an iterative process, to the cluster seed that is most similar to them. This procedure does not incorporate information about geographical regions. Thus, the countries assigned to a given cluster would not necessarily all be within the same geographical regions. A visual inspection determined that the additional subdivisions by geographical region would provide more homogeneous groups in terms of food consumption patterns.

At the second stage, the nine clusters were subdivided into subclusters based on a visual inspection of the appropriate geographical regions. A total of 13 subclusters were identified, based both on the similarity of the countries with respect to their consumption levels of the 19 marker foods and food groups, and on geographical proximity.

At the third stage, estimates were made of the intake of each of the 104 foods and food groups for each of the 13 cultural/regional diets. This calculation was made using the estimates of intake of each food by country (FAOSTAT.PC, Version 3.0, 1990–1994). If a country did not have a value for an individual food, it was assumed to be zero for these calculations.

The results of this cluster analysis are presented in Figure 8–2 and summarized below.

First Stage. Nineteen marker food groups were used in the cluster analysis, which summarizes the per capita consumption of these food groups for the nine statistical

124 INTERNATIONAL STANDARDS FOR FOOD SAFETY

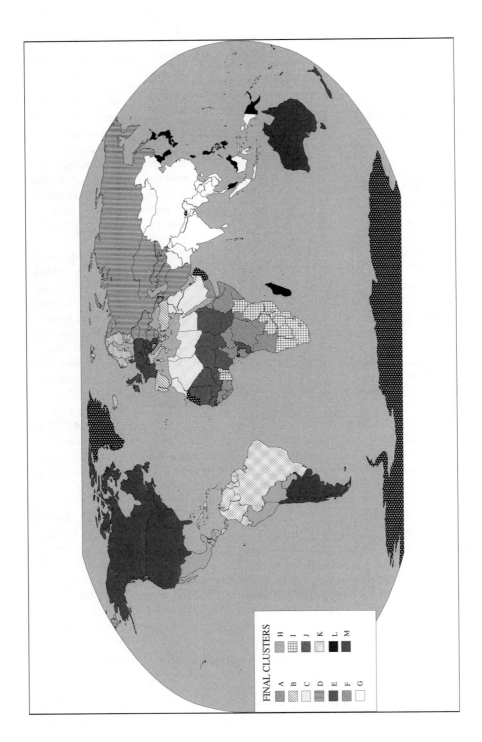

Figure 8–2 World Countries of Food Consumption

clusters. The clusters differ from each other with respect to their consumption level of one or more of these food groups. Cluster 1 identifies countries with relatively high maize consumption and relatively low wheat, meat, and fish consumption. Cluster 2 identifies countries with relatively high wheat and vegetable consumption and relatively low meat and fruit consumption. Cluster 3 identifies countries with relatively high wheat and meat consumption. Cluster 4 identifies countries with relatively high rice consumption. Cluster 5 identifies countries with high fruit and vegetable consumption and moderately high wheat consumption. Cluster 6 identifies countries with relatively high millet and sorghum consumption. Cluster 7 identifies countries with high fruit consumption. Cluster 8 identifies countries with high fish consumption, and Cluster 9 identifies countries with high wheat consumption and moderately high fruit and vegetable consumption.

Second Stage. The countries grouped under each of the statistical clusters did not all belong to the same geographical region, and may differ with respect to the consumption of other foods not included in the cluster analysis. In addition, some countries may have consumption patterns that could have them classified in other clusters.

Third Stage. The countries making each cluster were examined and divided further into subclusters based on geographical proximity. For instance, the countries making up Cluster 1 were, with a few exceptions, either African or South and Central American. The exceptions were Bosnia and Herzegovina and the Solomon Islands. A comparison of Bosnia's and Herzegovina's consumption levels with those of the countries making up Cluster 2 showed similar patterns. Thus, since countries making up Cluster 2 were essentially Eastern European, Bosnia and Herzegovina were moved with those countries. Similarly, a comparison of the Solomon Islands' consumption levels with those of the countries making up Cluster 8 showed similar patterns. Thus, Solomon Islands was moved with the countries making up this cluster. In addition, Sudan, Chad, and Nigeria had relatively high consumption levels of millet and sorghum, and were thus moved with the other African countries making up Cluster 6. Similar changes were made if necessary for the remaining clusters, thereby resulting in the identification of 13 subclusters, based on the primary statistical clusters and geographical regions.

There are some limitations in the use of the FBS's surveys to estimate intakes. First, waste at the household and individual levels usually is not considered. Therefore, intake estimates based on food supply data are higher than estimates based on actual food consumption survey data, with the magnitude of the error depending on the quantity of waste produced. Perhaps more importantly for intake assessment, users of foods cannot be distinguished from nonusers. Therefore, individual variations in intake cannot be assessed, nor can intake of potentially sensitive subpopulations be estimated. Finally, food availability is usually reported in terms of raw agricultural commodities. Processed forms of foods are usually not considered, nor is there any way to distinguish use of foods as ingredients. Nonetheless, these data allow assessments to be conducted at the international level using comparable data.

Household Inventories

Household surveys generally can be categorized as: (1) household or community inventories, or (2) household or individual food use. Inventories are accounts of what foods are available in the household. What foods enter the household? Were they purchased, grown, or obtained some other way? What foods are used up by the household? Were they used by household members, guests, and/or tenants? Were they fed to animals?

Inventories vary in precision with which data are collected. Questionnaires may or may not ask about forms of the food (i.e., canned, frozen, and fresh), source (i.e., grown, purchased, or provided through a food program), cost, or preparation. Quantities of foods may be inventoried as purchased, as grown, with inedible parts included or removed, as cooked, or as raw. Such data are available from many countries including Germany, the United Kingdom, Hungary, Poland, Greece, Belgium, Ireland, Luxembourg, Norway, and Spain (Trichopoulou & Lagiou, 1997).

Household or Individual Food Use

Food use studies, usually conducted at the household or family level, are often used to provide economic data for policy development and planning for feeding programs. Survey methods used include food accounts, inventories, records, and list-recalls (Pao et al., 1989; Lee & Nieman, 1993). These methods account for all foods used in the home during the survey period. This includes foods used from what was on hand in the household at the beginning of the survey period and foods brought into the home during the survey period.

Although household food use data have been used for a variety of purposes, including intake assessment, serious limitations associated with data from these surveys should be noted. Food waste often is not accounted for. Food purchased and consumed outside the household may or may not be considered. Users of a food within a household cannot be distinguished, and individual variation cannot be determined. Intakes by subpopulations based on age, gender, health status, and other variables for individuals can be estimated based only on standard proportions or equivalents for age/gender categories.

Individual Intake Studies

Individual intake studies provide data on food consumption by specific individuals. Methods for assessing food intakes of individuals may be retrospective (e.g., 24-hour or other short-term recalls, food frequencies, and diet histories), prospective (e.g., food diaries, food records, or duplicate portions), or a combination thereof. The most commonly used studies are those using the recall or record method and the food frequency. Each of the methods discussed below may also be applied at the family or household levels.

Recall Method (24-hour recall). Recalls are used to collect information on foods consumed in the past. The unit of observation is the individual or the household. The subject is asked to recall what foods and beverages he or she or the household consumed during a specific period, usually the preceding 24 hours. Since this method depends on memory, foods are quantified retrospectively, often with the aid of pictures, household measures, or two- or three-dimensional food models.

Recalls have been used successfully with individuals as young as six years of age, and interviewer-administered recalls are usually the method employed for populations with limited literacy or for individuals whose native language is not English. When individuals are not available for an interview or are unable to be interviewed due to age, infirmity, or temporary absence from the household, surrogate respondents are often used (Samet, 1989).

The main disadvantage of the recall method is the potential for error due to faulty memory of respondents. Items that were consumed may be forgotten, or the respondent may recall items consumed that actually were not consumed during the time investigated. To aid recall memories, the interviewer may probe for certain foods or beverages that are frequently forgotten, but this probing has also been shown to introduce potential bias by encouraging reporting of items not actually consumed.

Food Record/Diary Method. Records are used to collect information about current food intake. The subject is asked to keep a record of foods and beverages as they are consumed during a specific period. Quantities of foods and beverages consumed are entered in the record usually after weighing, measuring, or recording package sizes. Occasionally, photographs or other recording devices are employed.

Food Frequency Questionnaire Method. The food frequency questionnaire (FFQ), or checklist, determines the frequency of consumption of a limited number of foods, usually fewer than 100. This retrospective method estimates whether and how often foods are typically eaten. Subjects indicate how many times each day, week, or month they usually consume each food. Semiquantitative FFQs also estimate amounts consumed by allowing subjects to indicate whether their usual portion size is small, medium, or large compared to a stated medium portion.

Since FFQs collect data about intake of relatively few foods, a carefully tailored FFQ can provide useful data for assessment of chronic intake, particularly if the chemical in question is concentrated in only a few foods and if the food frequency instrument has been designed to target those foods.

Data from short-term recalls and from food diaries, which collect detailed information on the kinds and quantities of foods consumed, are generally the most accurate and flexible data to use in assessing intake of food chemicals. Data from these surveys can be used to estimate either acute or chronic intake. Averages and distributions can be calculated, and intake estimates can be calculated for subpopulations based on age, gender, ethnic background, socioeconomic status, and other demographic variables, provided that such information is collected for each individual.

In general, data from large surveys using recall/record methods provide the most accurate assessments of intake to chemicals in food when the chemical under investigation is found in many foods, and when most of these foods are consumed on a regular basis. It is more difficult to capture intake of infrequently consumed foods using short-term recalls or records.

Various survey methods are available for determining consumption of foods and beverages and where these items are consumed (e.g., food frequency interviews, food diaries, 24-hour dietary recalls, multiday weighed food record, and household food purchase records). For example, national dietary surveys were conducted in Australia in 1983 (adults) and 1985 (school children) and the entire population in 1995. Food consumption data are available from food frequency surveys for some states (ANZFA, 1997).

For the U.S. population, the data collected in one of two large, national, food-consumption surveys, conducted by the U.S. government, will be appropriate for most intake assessments in that country: (1) the Nationwide Food Consumption Survey conducted by the USDA, beginning in 1935 (USDA, 1987), and (2) the NHANES undertaken by the U.S. Department of Health and Human Services, beginning in 1971 (DHHS, 1996). Both surveys employ multistage area probability sampling procedures to obtain a sample representative of the population.

Special Situations That Require Unusual Data

Infrequently Consumed Foods

Some chemicals concentrate in liver and other foods that may be consumed infrequently. However, they may be eaten in significant quantities on those occasions when they are eaten. If an infrequently consumed food, such as liver, is a major source of the chemical under investigation, intake estimates for that chemical may be low. Intake assessments for those chemicals will be most accurate when based on surveys in which the frequency of consumption is estimated. In this case, a well-designed FFQ that targets the specific foods of concern should be considered.

Chemicals Found in Only a Few Foods

If the chemical is found in only a few foods, intake assessments for that chemical will be most accurate if the data used are from surveys that captured very specific information on foods consumed even when the survey does not capture information about the total diet.

Separating Sources and Consumption Locations of Food

If foods from one source, such as institutional foods, have different concentrations of chemicals than those from other sources, it will be appropriate to use food consumption data that permit exclusion of sources that do not contribute. The USDA CSFII (1989 to 1992) surveys provide extensive detail about the source of each food that each respondent reports consuming (USDA, 1992).

Foods That Are Unique to a Specific Population

Some subpopulations may consume foods that come from a unique source, and special steps are required to accurately assess these groups' intake. For example, one subpopulation consists of subsistence and recreational fishermen who consume fish that were caught from waters in which the chemical being studied is present at higher than typical levels. Another is vegetarians, whose diets are limited, to varying degrees, to nonanimal sources. Finally, infants and young children eat special foods and must be treated as a subpopulation. Commercial infant foods are prepared with special attention to the needs of infants and children, and manufacturers have programs to monitor environmental contaminants. Evaluation of data from the FDA monitoring programs indicates that levels will be lower for many chemicals. The presence of chemicals in breast milk should be evaluated in assessing nursing infants' intake of a chemical present in the food supply (Schreiber, 1997).

Validity and Reliability

Validity

Food consumption data used in assessing intake of chemicals in food should be both valid (accurate) and reliable (precise).

National consumption surveys and food monitoring programs are powerful tools for assessing dietary intake and evaluating status and trends. One must realize that these data are not intended to represent the true diets of all people in all places and that applying these data too broadly may result in false conclusions or poor estimates in dietary intake for specific populations. The factors affecting the intake of an individual or small subset of the population to a specific chemical are not evaluated by using a national average diet and nationwide residue monitoring data (Lioy, 1990). Market-basket surveys reveal little about unique, individual dietary practices that vary from the norm.

Numerous studies attempting to validate one survey method relative to another have been reported (Bingham et al., 1982; Blake & Durnin, 1963; Block, 1982; Bransby et al., 1948; Fanelli & Stevenhagen, 1986; Hussain et al., 1980; Karvetti & Knuts, 1985; Mahalko et al., 1985; Meredith et al., 1951; Morgan et al., 1987; Morrison et al., 1949; Nettleton et al., 1980; Ramsanen, 1979; Russell-Briefel et al., 1985; Willett et al., 1985; and Young et al., 1952). For example, the FFQ, a more recent survey method, has been validated by comparing results of dietary intake with repeated multiple-day food records, which served as an estimate of usual intake (Block, 1989; Pietinen et al., 1988a; and Willett et al., 1988). Results of validity studies indicated that FFQs could provide useful information about individual nutrient intakes. However, these studies generally have shown better correlations between methods for groups than for individual survey participants.

Validation of 24-hour recall, diaries, records, and frequencies has also been reported after comparing estimates of dietary intake obtained using one of these survey instruments to subjects' actual intakes. Actual intakes may be based on surreptitious observation in cafeterias, institutional dining centers, or other facilities (Baranowski et al., 1986; Gersovitz et al., 1978; Greger & Etnyre, 1978; Madden et al., 1976; Samuelson, 1970; and Stunkard & Waxman, 1981).

Survey methodology has also been validated by use of biological markers associated with dietary intake. Possible sources of biological markers include urine, feces, blood, and tissue samples, but the most easily accessible and therefore most commonly used is urine. Nitrogen content of urine has been used to verify protein intake. If protein intake calculated from the reported food intake is in agreement with protein intake calculated from nitrogen excretion, it is assumed that intake estimation for other nutrients is valid (Bingham & Cummings, 1985).

Reliability and Sources of Error

Reliability, or reproducibility, is the ability of a method to produce the same or similar estimate on two or more occasions (Block & Hartman, 1989; Pietinen et al., 1988b), whether or not the estimate is accurate. The reliability of food consumption survey data for estimating usual intake of a population depends somewhat on the number of days of dietary intake data collected for each individual in the population. The number of days of food consumption data required for reliable estimation of population intakes is related to each subject's day-to-day variation in diet (intraindividual variation) and the degree to which subjects differ from each other in their diets (interindividual variation) (Basiotis et al., 1987; Nelson et al., 1989). When intraindividual variation is small relative to interindividual variation, population intakes can be reliably estimated with consumption data from a smaller number of days than should be obtained when both types of variation are large. Intake of contaminants can be reliably estimated with fewer days of data when they are present in many foods that are commonly consumed.

In assessing food intake, it is generally accepted that mean intake of a population may be reasonably estimated using a one-day recall or diary if the number of subjects is sufficiently large. However, the percentage of the population estimated to be at risk for toxic effects from a chemical will be higher when food intake is assessed using a one-day recall than with a multiday record or dietary history. This is because extreme levels of intake (e.g., 90th or 95th percentiles) are invariably higher for a single day than they are for multiple days. In addition, large intraindividual variation associated with one-day surveys may limit the power to detect differences between different population groups (Liu et al., 1978; Beaton et al., 1979, and van Staveren et al., 1985).

Error in individual food consumption surveys may be due to chance or to factors of measurement. Data variability due to chance may be related to the survey sample. Any sample randomly drawn from a population will differ from any other sample, with the degree of difference depending on the size of the sample and the homogeneity of the population from which it was drawn. Error due to chance also arises from data collection at different times of the day, on different days of the week, or in different seasons of the year.

Measurement error may be introduced by the survey instrument, the interviewer, or the respondent. The instrument may bias results when questions are not clear, probes lead the subject to give a desired answer, questions are culture-specific, or they do not follow a logical sequence. For self-administered questionnaires, responses will be influenced by the readability level, the use of abbreviations or unfamiliar jargon, clarity of instructions, and amount of space provided for answers. Interviewer bias may be introduced when interviewers make the respondent uncomfortable, are judgmental, or do not use a standard method and/or standard probes.

Respondents may introduce bias when they omit reporting foods they actually ate because they are reluctant to report certain foods or beverages (alcoholic beverages are a good example) or if they are forgetful. Alternatively, they may report the food but understate the quantity consumed. Foods consumed away from home, particularly on occasions when the focus of attention is on the event rather than on the food, are especially difficult for people to remember. Quantities may be underestimated for similar reasons. Foods and beverages that were not consumed may be reported as consumed because of faulty memories, desire to impress the interviewer, or confusion with similar foods.

Measurement errors also include errors in coding due to unclear handwritten records or erroneous data entry.

LIMITATIONS AND NEED FOR RESEARCH

Short-Term Versus Long-Term Intake Needs To Be Differentiated

All analyses should be relevant for the specific chemical and take advantage of the available data. The available data represent a snapshot of dietary intake representing at most a week of food intake records. For many purposes, long-term intake patterns are desirable. Methods are thus needed to estimate dietary intake for longer time periods including up to a year or more.

Food Additives' Levels in Foods as They Are Consumed

The concentrations of most food additives in foods are not well quantified. Maximum permitted levels may be found, but when analytical measurements have been made, the levels found were substantially lower. Further, many foods contain no additives. In most cases, the food industry regards this as highly proprietary information. Methods need to be devised to permit obtaining and using the information while at the same time protecting industry proprietary information.

Impact of Food Processing

Most additives are added early in the preparation of foods. The ingredients then undergo additional processing. The impact of this processing on the food additive needs to be better understood.

Improving Correlation between Countries

There are many situations where it is desirable to determine intake by individuals in different countries. The work done by the experts at the FAO/WHO conference (FAO/WHO, 1998) demonstrated the significant differences in dietary practices.

Thirteen major regional diets were identified. Within those regions, dietary patterns can be further subdivided. The available data for defining these differences are extremely limited. Further, it is extremely difficult to compare estimates between countries. Such comparisons are needed to guide international standard setting (Codex Alimentarius Commission [CAC, 1998]) and to facilitate food safety evaluations.

CONCLUSION

The validity of any intake calculation depends upon the representativeness and reliability of the available data. It also depends upon appropriate selection and use of the algorithms and models. Independent validation and sensitivity analyses to quantify the uncertainty of the estimates are critical components.

REFERENCES

Australia New Zealand Food Authority. (1997). *Dietary Modelling: Principles and Procedures.* Canberra, Australia: Australia New Zealand Food Authority.

Baranowski, T., Dworking, R., Henske, J. C., Clearman, D. R., Dunn, J. K., Nader, P. R. & Hooks, P. C. (1986). The accuracy of children's self-reports of diet: family health project. *J Am Diet Assoc* 86, 1380.

Barraj, L., Petersen, B. J. & Moy, G. (1999). Creation of dietary regions using cluster analysis. In press.

Basiotis, P. P., Welsh, S. O., Cronin, J., Kelsay, J. L. & Mertz, W. (1987). Number of days of food intake records required to estimate individual and group nutrient intakes with defined confidence. *J Nutr* 117, 1638.

Beaton, G. H., Milner, J., Corey, P., McGuire, V., Cousins, M., Stewart, E., de Ramos, E., Hewitt, D., Grambsch, P. V., Kassim, N. & Little, J. A. (1979). Sources of variance in 24-hour dietary recall data: implications for nutrition study design and interpretation. *Am J Clin Nutr* 32, 2456.

Bingham, S. & Cummings, J. H. (1985). Urine nitrogen as an independent validatory measure of dietary intake: a study of nitrogen balance in individuals consuming their normal diet. *Am J Clin Nutr* 42, 1276.

Bingham, S., Wiggins, H. S., Englyst, H., Seppanen, R., Helms, P., Strand, R., Burton, R., Jorgensen, I. M., Poulsen, L., Paerregaard, A., Bjerrum, L. & James, W. P. (1982). Methods and validity of dietary assessments in four Scandinavian populations. *Nutr Cancer* 4, 23.

Blake, E.C. & Durnin, J. V. G. A. (1963). Dietary values from a 24-hour recall compared to a 7-day survey on elderly people. *Proc Nutr Soc* 22, 1.

Block, G. (1982). A review of validations of dietary assessment methods. *Am J Epidemiol* 115, 492.

Block, G. (1989). Human dietary assessment: methods and issues. *Prevent Med* 18, 653.

Block, G. & Hartman, A. M. (1989). Issues in reproducibility and validity of dietary studies. *Am J Clin Nutr* 50, 1133.

Bolger, P. M., Yess, N. J., Gunderson, E. L., Troxell, T. C. & Carrington, C. D. (1996). Identification and reduction of sources of dietary lead in the USA. *Food Addit Contam* 13, 53–60.

Bransby, E. R., Daubney, C. G. & King, J. (1948). Comparison of results obtained by different methods of individual dietary survey. *Br J Nutr* 2, 89.

Chin, H. B. (1991). The effect of processing on residues in foods: the food processing industry's residue database. In *Pesticide Residues and Food Safety: A Harvest of Viewpoints*, p. 24. Edited by B. G. Tweedy, H. J. Dishburger, L. G. Ballantine, and J. McCarthy. Washington, DC: American Chemical Society.

Codex Alimentarius Commission. (1998). *Report of the 30th Session of the Codex Committee on Food Additives and Contaminants.* ALINORM 99/12, Rome: Codex Alimentarius Commission.

Dabeka, R. W., McKenzie, A. D. & Lacroix, G. M. A. (1987). Dietary intakes of lead, cadmium arsenic and fluoride by Canadian adults: a 24-hour duplicate-diet study. *Food Addit Contam* 4, 89–102.

Fanelli, M. T. & Stevenhagen, K. J. (1986). Consistency of energy and nutrient intakes of older adults: 24-hour recall vs. 1-day food record. *J Am Diet Assoc* 86, 664.

Food and Agriculture Organization/World Health Organization. (1995). *Recommendations for the Revision of Guidelines for Predicting Dietary Intakes of Pesticide Residues.* Report of a FAO/WHO Consultation, May 2–6, 1995, York, United Kingdom. WHO/FNU/FOS/95.11. Geneva, Switzerland: World Health Organization.

Food and Agriculture Organization/World Health Organization. (1998). *Food Consumption and Exposure Assessment of Chemicals.* Report of a FAO/WHO Consultation, February 10–14, 1997, Geneva. WHO/FSF/FOS/97.5. Geneva, Switzerland: World Health Organization.

Federation of American Societies of Experimental Biology. (1988). *Estimation of Exposure to Substances in the Food Supply.* Bethesda, MD: Life Sciences Research Office.

Federation of American Societies of Experimental Biology. (1993). *National Human Exposure Assessment Survey Dietary Monitoring Options.* Bethesda, MD: Life Sciences Research Office.

Gersovitz, M., Madden, J. P. & Smiciklas-Wright, H. (1978). Validity of the 24-hour dietary recall and seven-day record for group comparisons. *J Am Diet Assoc* 73, 48.

Graham, J., Walker, K., Berry, M., Bryan, E., Callahan, M., Fan, A., Finley, B., Lynch, J., McKone, T., Ozkaynak, H. & Sexton, K. (1992). The role of exposure data bases in risk assessment. *Arch Environ Health* 47, 408–420.

Greger, J. L. & Etnyre, G. M. (1978). Validity of 24-hour recalls by adolescent females. *Am J Pub Health* 68, 70.

Hansen, S. C. (1979). Conditions for use of food additives based on a budget for an acceptable daily intake. *J Food Protect* 42, 429.

Hussain, M. A., Abdullah, M., Huda, N. & Ahmad, K. (1980). Studies on dietary survey methodology—a comparison between recall and weighing method in Bangladesh. *Bangladesh Med Res Counc Bull* 6, 53.

Karvetti, R. L. & Knuts, L. R. (1985). Validity of the 24-hour recall. *J Am Diet Assoc* 85, 1437.

Lee, R. D. & Nieman, D. C. (1993). *Nutritional Assessment.* Dubuque, Iowa: William C. Brown Publishers.

Lioy, P. J. (1990). Assessing total human exposure to contaminants. *Environ Sci Technol* 24, 938–945.

Liu, K., Stamler, J., Dyer, A., McKeever, J. & McKeever, P. (1978). Statistical methods to assess and minimize the role of intra-individual variability in obscuring the relationship between dietary lipids and serum cholesterol. *J Chron Dis* 31, 399.

Madden, J. P., Goodman, S. J. & Guthrie, H. A. (1976). Validity of the 24-hr. recall. *J Am Diet Assoc* 68, 143.

Mahalko, J. P., Johnson, L. K., Gallagher, S. K. & Milne, D. B. (1985). Comparison of dietary histories and seven-day food records in a nutritional assessment of older adults. *Am J Clin Nutr* 42, 542.

Meredith, A., Matthews, A., Zickefoose, M., Weagley, E., Wayave, M. & Brown, E. G. (1951). How well do school children recall what they have eaten? *J Am Diet Assoc* 27, 749.

Morgan, K. J., Johnson, S. R., Rizek, R. L., Reese, R. & Stampley, G. L. (1987). Collection of food intake data: an evaluation of methods. *J Am Diet Assoc* 87, 888.

Morrison, S. D., Russell, F. C. & Stevenson, J. (1949). Estimating food intake by questioning and weighing: a one-day survey of eight subjects. *Proc Nutr Soc* 7, v.

National Center for Health Statistics. 1982. *Second National Health and Nutrition Examination Survey. 1976–80.* Data Tape 5704 (24-Hour Recall, Specific Food Item), NTIS Accession number PB82–142639. Springfield, Virginia: National Technical Information Service.

Nelson, M., Black, A. E., Morris, J. A. & Cole, T. J. (1989). Between- and within-subject variation in nutrient intake from infancy to old age: estimating the number of days required to rank dietary intakes with desired precision. *Am J Clin Nutr* 50, 155.

Nettleton, P., Day, K. C. & Nelson, M. (1980). Dietary survey methods. 2. A comparison of nutrient intakes within families assessed by household measures and the semi-weighed method. *J Hum Nutr* 34, 349.

Pao, E. M., Sykes, K. E. & Cypel, Y. S. (1989). *USDA Methodological Research for Large-Scale Dietary Intake Surveys. 1975–88.* Home Economics Research Report No. 49. Washington, DC: U.S. Department of Agriculture, Human Nutrition Information Service.

Pennington, J. A. T. & Gunderson, E. T. (1987). History of the Food and Drug Administration's Total Diet Study—1961 to 1987. *J Assoc Off Anal Chem* 70, 772–782.

Petersen, B. & Barraj, L. (1996). Assessing the intake of contaminants and nutrients: a review of methods. *J Food Composition Anal* 9, 243–254.

Petersen, B. & Douglass, J. (1994). Use of food-intake surveys to estimate exposure to nonnutrients. *Am J Clin Nutr* 59(Suppl.), 2403–2433.

Pietinen, P., Hartman, A. M., Haapa, E., Rasanen, L., Haapakoski, J., Palmgren, J., Albanes, D., Virtamo, J. & Huttenen, J. K. (1988a). Reproducibility and validity of dietary assessment instruments. I. A self-administered food use questionnaire with a portion size booklet. *Am J Epidemiol* 128, 655.

Pietinen, P., Hartman, A. M., Haapa, E., Rasanen, L., Haapakoski, J., Palmgren, J., Albanes, D., Virtamo, J. & Huttenen, J. K. (1988b) Reproducibility and validity of dietary assessment instruments. II. A qualitative food frequency questionnaire. *Am J Epidemiol* 128, 667.

Ramsanen, L. (1979). Nutrition survey of Finnish rural children. VI. Methodological study comparing the 24-hour recall and the dietary history interview. *Am J Clin Nutr* 32, 2560.

Romesburg, H. C. (1990). *Cluster Analysis for Researchers.* Malabar, FL: Krieger Publications.

Russell-Briefel, R., Caggiula, A. W. & Kuller, L. H. (1985). A comparison of three dietary methods for estimating vitamin A intake. *Am J Epidemiol* 122, 628.

Samet, J. M. (1989). Surrogate measures of dietary intake. *Am J Clin Nutr* 50, 1139.

Samuelson, G. (1970). An epidemiological study of child health and nutrition in a northern Swedish county. 2. Methodological study of the recall technique. *Nutr Metab* 12, 321.

Sasaki, S. & Kestelhoot, H. (1992). Value of Food and Agriculture Organization data on food-balance sheets as a data source for dietary fat intake in epidemiologic studies. *Am J Clin Nutr* 56, 716.

SCF (Intake and Exposure Working Group). (1994). *Summaries of food consumption databases in the European Union.* CS/Int/Gen2. Brussels, Belgium: European Commission.

Schreiber, J. S. (1997). Transport of organic chemicals to breast milk: tetrachloroethene case study. In *Environmental Toxicology and Pharmacology of Human Development,* pp. 95–143. Edited by S. Kacew and G. Lambert. Washington, DC: Taylor and Francis.

Sexton, K., Selevan, S. G., Wagner, D. C. & Lybarger, J. A. (1992). Estimating human exposures to environmental pollutants: availability and utility of existing databases. *Arch Environ Health* 47, 398–407.

Sherlock, J. C., Smart, G. A., Walters, B., Evans, W. H., McWeeny, D. J. & Cassidy, W. (1983). Dietary surveys on a population at Shipham, Somerset, United Kingdom. *Sci Total Environ* 29, 121–142.

van Staveren, W. A., de Boer, J. O., and Burema, J. (1985). Validity and reproducibility of a dietary history method estimating the usual food intake during one month. *Am J Clin Nutr* 42, 554.

Stunkard, A. J. & Waxman, M. (1981). Accuracy of self-reports of food intake. *J Am Diet Assoc* 79, 547.

Trichopoulou, A. & Lagiou, P., eds. (1997). *Methodology for the exploitation of HBS Food Data and Results on Food Availability in 5 European Countries.* DAFNE (Data Food Networking) European Communities.

U.S. Department of Agriculture. (1987). Nationwide Food Consumption Surveys of Food Intake by Individuals and the Continuing Survey of Food Intake by Individuals (1989–1994). Springfield, VA: National Technical Information Services.

U.S. Department of Agriculture. (1991 and 1996). USDA Continuing Survey of Food Intake by Individuals (CSFII), 1989–91 and 1994–96. Springfield, VA: National Technical Information Services.

U.S. Department of Agriculture. (1992). USDA CSFII surveys (1989–1992). Springfield, VA: National Technical Information Services.

U.S. Department of Health and Human Services. (1996). Third National Health and Nutrition Examination Survey (1988–1994). Springfield, VA: National Technical Information Services.

Vahter, M., Berglund, M., Friberg, L., Jorhem, L., Lind, B., Slorach, S. & Akesson, A. (1990). Dietary intake of lead and cadmium in Sweden. *Var Foda* 44 (Suppl. 2). Uppsala, Sweden: National Food Administration.

Willett, W. C., Sampson, L., Stampfer, M. J., Rosner, B., Bain, C., Witschi, J., Hennekens, C. H. & Speizer, F. E. (1985). Reproducibility and validity of a semi-quantitative food frequency questionnaire. *Am J Epidemiol* 122, 51.

Willett, W. C., Sampson, L., Browne, M. L., Stampfer, M. J., Rosner, B., Hennekens, C. H. and Speizer, F. E. (1988). The use of a self-administered questionnaire to assess diet four years in the past. *Am J Epidemiol* 127, 188.

World Health Organization. (1983). *Assessment of Human Exposure to Environmental Pollutants.* EFP/83.52. Geneva, Switzerland: World Health Organization.

World Health Organization. (1988). *Derived Intervention Levels for Radionuclides in Food.* Geneva, Switzerland: World Health Organization.

World Health Organization. (1997). Guidelines for Predicting Dietary Intake of Pesticide Residues. GEMS/Food. Geneva: World Health Organization.

World Health Organization. (1998). *GEMS/Food Regional Diets, Regional per Capita Consumption of Raw and Semi-Processed Agricultural Commodities.* WHO/FSF/FOS 98.3. Geneva, Switzerland: World Health Organization.

Young, C. M., Hagan, G. C., Tucker, R. E. & Foster, W. D. (1952). A comparison of dietary study methods. II. Dietary history vs. seven-day record vs. 24-hour recall. *J Am Diet Assoc* 28, 218.

Chapter 9

The Need for Developing Countries To Improve National Infrastructure To Contribute to International Standards

Lillian T. Marovatsanga

INTRODUCTION

In the past several decades, all nations have experienced extensive changes and are faced with new challenges in ensuring adequate national food security and safe and high-quality foods for international trade and consumer health. As a result, food control programs based on well-established food standards and other regulations are more necessary now than ever before to improve nutritional status, control food quality and safety, prevent food losses, and protect the consumer. Effective food standards and control systems are required not only to protect national food production, but also to facilitate trade within the national borders and with other nations. Food control and standards are the common threads that bind together food security, international trade, and consumer health in a practical, meaningful way.

In anticipation of the greater importance of food standards in international trade and the role to be played by the Codex Alimentarius Commission (CAC) in the food standards-setting process, a major conference on Food Standards, Chemicals in Food and Food Trade was held in Rome, March 1991. One of the major recommendations of the conference was harmonization of national food regulations in order to facilitate international trade. The need to improve food quality, harmonize national food requirements with international standards, and institute adequate food control systems in developing countries was considered of the highest importance by the attendees of the conference.

Lillian T. Marovatsanga tragically passed away during the production of this book. John Kapito, executive director of the Consumers Association of Malawi, kindly reviewed Dr. Marovatsanga's chapter for publication. The book editors wish to dedicate this chapter to her.

Achieving technical harmonization through international trade agreements requires further concentrated work at the international, regional, and national levels. The major constraint in developing countries that prohibits their full contribution and participation in international standards setting is lack of infrastructure. Food legislation and enforcement, encompassing food standards and codes of hygiene practice, are areas that should be assessed by government authorities in order to identify major problems of control relevant to food safety. Also, food legislation is often not reviewed and updated. Hence national standards are always lagging behind, a situation that should be changed by improving the national infrastructure so that developing nations become more involved in international standard setting.

Effective application of any standard requires legislation and enforcement. It further requires adequately equipped and trained staff performing inspectorate and laboratory functions and a compliance program developed from a well-thought-out food control strategy applied both to domestic and import/export food production. Unfortunately, many developing countries with economies in transition lack the necessary food control infrastructure to implement both strategy and programs to protect their consumers and to overcome their disadvantaged status in the international trade in food. These problems include:

- inadequate or outdated legislation and regulations
- inadequate resources and/or failure to maximize available resources
- failure to develop a national food control strategy and poorly implemented or managed program and activities
- inadequately equipped laboratories and inspectorate
- inadequately trained and technically deficient personnel
- poor coordination and cooperation among food control agencies, other concerned government agencies, academia, industry, and consumers in order to suit international standard setting
- lack of political will and commitment on food safety and standards by developing countries

Lack of expertise and financial resources are major problems for many developing countries. Poor administration of available resources adds to the problems.

In a national food control program, there is a need to develop and update food legislation as well as train food handlers, inspectors, analysts, and consumers. Also research and data collection work should be conducted on food laws and regulations, sample collection, and testing methods. Due to globalization of markets, it is becoming increasingly important to use international standards. The work of international organizations in food control, such as the Food and Agriculture Organization and World Health Organization (FAO/WHO), should be used as ideal models to create a national standards body that would regulate and enforce the standards, guidelines, and recommendations of the Joint FAO/WHO CAC.

Any national body should include representatives from government agencies, the food manufacturing sector, the academia, and consumers—all have different responsibilities in food safety. The primary role of the government is to provide leadership for the implementation of standards. Through food control agencies, the government can advise the food industry on the benefits of implementing voluntary standards. Expert working groups and national seminars and workshops would also assist in creating national standards. With the active support of the government, the body can encourage the application of standards by industrialists to improve their competitive edge in the international market.

Some of the activities of a national standards body are as follows:

- formulation of draft standards and publishing standards
- quality control and testing—to make available to producers and consumers, laboratory facilities for testing of manufactured goods and raw materials
- certification—to operate a product certification scheme whereby manufacturers apply for their products to be certified against existing product standards

The principles underlying implementation of national standards should be national uniformity, international best practice, appropriate but minimum food regulations, cost-effectiveness, contestable compliance, and consistency in approach throughout the food supply chain.

INTEGRATED APPROACH TO STANDARDIZATION

An integrated approach is important to cover all technical fields such that the practice of standardization becomes better coordinated.

Levels of Standardization

There are four levels of standards:

1. **Company standards**—These are prepared by a company for its own use. It is common to find company standards that are copies of national standards.
2. **National standards**—These are issued by the national standards body.
3. **Regional standards**—Regional groups with similar geographical, climate, and cultural factors have legislation standardization bodies (e.g., African Regional Organization for Standardization).
4. **International standards**—The International Organization for Standardization (ISO) and Codex Alimentarius publish international standards.

While standards can be considered at these four different levels, there is coordination among them. Company standards are used in formulating national standards; likewise national standards provide information for regional and international standards and vice versa.

NATIONAL FOOD CONTROL: AN OUTLINE OF THE MAJOR CONSTRAINTS

The major constraints of national food control are:

- insufficient funding for food control activities such as inspection, sampling, testing, and analysis
- inadequate manpower for inspection and analysis, especially in government departments (this has resulted in poor enforcement of legislation, especially in rural areas and an influx of substandard imported foods)
- poor participation and representation at international/national food standards conferences, workshops, and seminars by developing countries
- absence of appropriate in-service training, especially in areas of food legislation and inspection
- inadequate facilities and infrastructure for food analysis
- consumer ignorance on issues of food quality and food handling, as well as the absence of consumer education programs

Pertaining to food standards and regulations, there are the following constraints:

- There are too many government regulations dealing with food that are administered by different ministries.
- Voluntary standards produced by private standard-setting bodies can be incorporated into government regulations and become mandatory.
- Participation in Codex work has been hampered by lack of funds and the absence of recognized national Codex committees.
- Lack of technology and expertise has caused delays in adoption of some standards by some developing countries.

AREAS NEEDING IMMEDIATE ATTENTION TO STRENGTHEN NATIONAL FOOD CONTROL INFRASTRUCTURE

Developing and Updating Food Legislation

Legislation requires continuous review, and new regulations on food export and food aid should be developed. Food laws and regulations should be sufficiently flexible while maintaining minimum standards to meet the needs of a changing food sector, the introduction of modern technology, and the development of new food products.

National governments and major world trading blocks are using food safety risk assessment, risk management, and risk communication to improve food standards. Risk management of contaminants in food requires assessment of dietary exposure. National estimates of dietary exposure should, as far as feasible, be consistent with approaches used at the international level. Uncertainty in dietary exposure assessments may require national authorities to consider the intake models developed at the international level and to identify major discrepancies or omissions.

The limitations of developing countries in their ability to perform comprehensive exposure assessments are as follows:

- lack of databases required as well as the infrastructure (i.e., the technical resources to utilize these databases)
- lack of necessary expertise
- lack of detailed international information on how to perform risk analysis

The subject of food quality is seen as a matter principally for the food industry to ensure thorough self-regulation. Improvement of quality standards within the food industry of any given country has three major implications: (a) for food manufacturers (considering that there is legislation), it reduces the cost due to outbreaks and incidents; (b) for operators of the food chain, it increases competitiveness and income generated from sales; and (c) for the population and for the government, it is directly linked to the five above-mentioned mechanisms used by national authorities to ensure the availability of a safe variety of food supply. Improvement of quality standards is linked to food safety through the establishment of and compliance with legislation, its enforcement, and the identification of where and what type of possible contamination and deviation from the quality standards is occurring. Therefore, government's role then becomes that of inspecting and verifying that industry management systems for food quality assurance are working effectively and complying with set standards.

Owing to current trends in harmonization of regional and international standards caused by the requirements for international trade, a framework food safety law should be set up nationally to encourage the strengthening of legislation. This law would:

- Define methods to enforce legislation and identify penalties for breaches of the law.
- Provide a mechanism for the introduction of subsidiary legislation and specific regulations, such as codes of practice, that will contain specific details on such matters as enforcement procedures, regulations on hygiene, use of food additives and labeling, licensing of food premises, and import or export regulations.

Developing and updating food legislation is a complex task that needs input from many disciplines including food safety experts, the legal profession, the academia, industry and consumer group representatives, and the government. Extensive consultation with the food sector and consumer groups, which is important in reviewing national and international food legislation, is facilitated through the organization of workshops, seminars, conferences, and working groups.

A Growing National Stake in Standards Issues

The government should have a considerable interest in the effectiveness of the standard-setting process. Standards help determine the efficiency and effectiveness of the economy, as well as the cost, quality, and availability of products and services.

Support by the government in applying standard obligations improves the state of the nation's health, safety, and quality of life.

Failure by the government and standards development organizations to appreciate the implications of international standards will have serious consequences for industry. The global market is opening up, and there is intense competition demanding quality products (ISO, 1996). For example, the Japanese have gained considerable ground in the international market, partly by more effectively using standards to improve productivity and add value to their products.

Members of the national standards body should consult international developments and standards in order to develop their own standards based on:

- industrial development in the country
- level of technological advancement
- availability of sound technological and scientific bases in the country
- volume of trade (both import and export)

Strengthening Food Control Systems

In order to enforce the adoption and implementation of set standards, there has to be a well-integrated, nationally coordinated food control system. An effective food control system must consist of an administration (central or local), an inspectorate, and analytical capability. Centralized administration has been shown to be a good model for an effective food control program. Formation of a national food control administration would lead to a more efficient and cost-effective system.

Food Inspection

Conformity of products to standards is verified through inspection, which must verify that all foods are produced, handled, processed, packed, stored, and distributed in compliance with regulations and legislation. Responsible ministries and municipal authorities need to assign people to investigate the status of hygienic quality and conformity with standards of street-vended food, food-handling practices of street vendors, and all their facilities. A strategic plan for correction measures should be developed when the results show nonconformity to set standards.

Analytical Facilities

There is a need for more laboratories and expansion of the existing facilities. To improve the laboratory services, modern equipment and more trained personnel are a prerequisite.

A broad range of analytical capabilities is required for detecting food contaminants such as pesticides, pathogenic bacteria, food-borne viruses and parasites, radionuclides, environmental chemicals, and biotoxins. Capability is also required to determine food adulteration and compliance with official food quality standards (see *Manual of Food Quality*).

Laboratory services need to be decentralized. They should be established at provincial level and border posts, for important basic tests on imported foods. This would form a strong infrastructure for facilitating implementation of international standards.

Operational procedures in the laboratories should conform to the internationally recognized guidelines specified in the ISO/IEC Guide 25, detailing the General Requirements for the Competence of Calibration and Testing Laboratories Food Safety Issues for international acceptance of the competence of methods and results obtained.

Certification and registration activities rely on information obtained from laboratory tests to confirm compliance of products to standards. Therefore, absence of the appropriate testing equipment enables manufacturers to compromise on quality since they know that their products will not be tested due to lack of equipment. In international markets, however, this tactic could result in the rejection of big consignments, which could be very costly to the manufacturer.

Training

All personnel in food control services should be aware of the requirements and problems associated with the production and regulations of a safe and nutritious food supply. With adequate allocation of resources, appropriate training programs of these personnel can help achieve this goal. Both the food industry and government should be actively involved in the organization, support, and implementation of training programs if they are to be effective. Such programs can achieve their aims if they are supported, designed, developed, and implemented by health workers, food safety authorities, and consumer groups.

Incorporation of standards in college curricula would be a step in the right direction toward training people. Joint seminars, workshops, and conferences also can be used in training. Relevant topics for all forms of training include:

- introduction of the hazard analysis critical control point (HACCP) system
- principles of auditing and verification of the implemented HACCP systems in industry to check if they ensure an acceptable level of prevention and control of risks, and that programs are being carried out effectively
- principles of laboratory management to comply with the ISO/IEC Guide 25, and in microbiological, chemical, and physical analysis of foods

The food industry sector should initiate programs to train food handlers in the principles of food hygiene—relating nature of food and its ability to sustain growth and survival of pathogenic microorganisms to risks of contamination during food handling. Training should also focus on the principles of good agricultural practice, good manufacturing practice, and the principles and application of voluntary quality assurance system based on HACCP (WHO, 1996).

Education of Handlers and Consumers

Educating food handlers (e.g., food processing, manufacturing, catering, and domestic handlers) to adhere strictly to good personal hygiene and to provide hygienic food shows that handling practices are very important. Handlers can cause contamination through inadequate personal hygiene or when handling food when they are medically unfit. Food contamination causes food poisoning, which can have serious consequences ranging from illness to death. Food contamination may lead to survival and multiplication of pathogenic microorganisms (e.g., *Salmonella*, *Escherichia coli,* and *Staphylococcus aureus*) to disease-causing levels, which are above stipulated standards.

Appropriate education programs on domestic food preparation and storage are needed, particularly in developing countries, where access to the food cold chain is not always possible, and foods are prepared and stored at high ambient temperatures. As a result, if the food is contaminated, there is a risk of multiplication of pathogenic bacteria, thereby posing a health risk.

Consumers need to be educated about food standards. These standards can be disseminated through the mass media. In addition, information services need to be improved to make it easier to access data. Computerized data-keeping systems make retrieving information more efficient. New information technology such as e-mail should be readily available so that consumer organizations and enforcement agents can exchange information easily.

Coordination of Research Work

Collaboration among universities, other research institutions, government laboratories, and private laboratories is important in order to avoid duplication of effort, which is not cost effective. In addition, sharing of ideas via an interlaboratory quality system would help with the use of the uniform testing methods recommended by internationally recognized associations.

Reliable information is necessary when setting priorities for allocation of resources, for consumer confidence in the food supply, and for supporting trade, particularly national exports. For standards to be applied in a way that does not lead to unnecessary trade barriers, they have to be based on scientific data and knowledge. Governments require objective and sound scientific advice on food-related problems, which could be obtained from a number of independent expert committees or working groups. Ways in which this can be done include:

- formation of advisory committees on such topics as the microbiological safety of foods, toxicity of chemicals in foods, novel foods and processes, and on pesticides and veterinary products
- formation of a steering group on food surveillance to keep under review the possibility of contamination of any part of the national food chain and to report to responsible ministries

- formation of a group of experts that recommends that the national diet is both safe and nutritious based on local needs and compliance with international standards
- formation of effective national Codex committees

The Role of Industry

Industry should also have the necessary skills through appropriately trained, knowledgeable personnel. Application of new technology requires the use of up-to-date machinery for manufacturing. Using equipment in good condition eliminates the problems of producing substandard products due to faulty equipment. Equipment should be accurate, reliable, and sufficient in quality and variety to meet requirements, as well as wherever possible, of the latest design.

Industrial purchasers can seek information and advice from bulk purchasing agencies before buying equipment to avoid obtaining deficient products. Doing so is an important part of applying standards.

The industry should also expand its efforts to develop innovative ways to share information and foster global discussions on new technologies and development in food production and processing, which includes information and data relevant to form acceptable international and national standards.

In cooperation with FAO/WHO, the food industry can promote the use of the best practices and technology transfer to achieve enhanced and sustainable food quality.

Approaches to the Application of Standards

All the standards in critical areas affecting the health and safety of the nation should be made compulsory to ensure that all available goods always comply with the requirements. In this way, the benefits of the standardization process would be directed to meet the needs of the public. The sufficient provision of raw materials, appropriate equipment for production and testing, skilled personnel, and all other required resources, however, is key to the success of a mandatory approach.

Industry should also be encouraged to adopt voluntary standards. Demand can be created through publicity. Benefits of adopting national/international standards, such as opening up trade and ability to face competition, should be made clear to manufacturers.

Benefits of Enforcing Adoption of the National Standards

There are many benefits to adopting national standards. These include:

- proper utilization of scarce national resources
- increased productivity through rationalization of varieties

- consumer protection by making available products of standard quality
- ensuring safety from industrial and household hazards
- ensuring safe nutritional levels and freedom from toxicity in food products
- efficient utilization of energy resources
- controlling hazardous effluent, pollutants, and emissions (by using an information manual)

EVALUATION AND ASSESSMENT OF FOOD CONTROL AND SAFETY PROGRAMS

Evaluation is a necessary stage in project management, and it follows from the planning stage through to implementation and monitoring. In developing countries, the national authorities face problems that make the development, implementation, and maintenance of effective and efficient food safety and control programs difficult to achieve. These problems are caused by such factors as constraints on the availability of human and financial resources and inability to respond to rapid technological changes in the production, processing, and distribution of foods.

Important factors that should be considered during evaluation of a national food control program are effectiveness, impact, efficiency, progress, adequacy, and relevance. Then the results of the evaluation exercise need to be objectively interpreted. Results can contribute to strengthening the program.

SUMMARY

The Codex Coordinating Committee for Africa has advised that:

- National food laws and regulations for food import and export inspection and certification should be harmonized with existing Codex standards and guidelines and in consideration of the Application of Sanitary and Phytosanitary Measures and the Agreement on Technical Barriers to Trade.
- Government should give the utmost priority and support to the strengthening of national food control activities by providing increased resources, including support to allow the participation of government representatives at Codex committee sessions.
- National intra-agency, intergovernmental coordination and harmonization should be rationalized to allow for maximum efficiency, utilization of resources, and exchange and dissemination of information to the public and private sectors.
- International bodies and governments should be encouraged to provide assistance to developing countries in any way possible.

BIBLIOGRAPHY

Codex Alimentarius Commission. (1985). *Report of the 16th Session of the Joint FAO/WHO Codex Alimentarius Commission Held in Geneva.* Rome: Codex Alimentarius Commission.

Codex Alimentarius Commission. (1994). *Joint FAO/WHO Food Standards Programme: Codex Committee on General Principles*. 11th Session, Paris, April 25–29, 1994. Rome: Codex Alimentarius Commission.

Codex Alimentarius Commission. (1995). *Preamble to the Draft Codex General Standard for Contaminants and Toxins*. ALINORM 95/12A, Appendix IV. Rome: Codex Alimentarius Commission.

Codex Alimentarius Commission. (1996). *Codex Alimentarius Manual*, 2nd ed. Rome: Food and Agriculture Organization.

Codex Committee on Food Additives and Contaminants. (1996). *Codex Risk Assessment and Management Procedures: Method To Ensure Public Safety While Developing Codex General Standard for Contaminants and Toxins in Food*. CX/FAC 96/15. Rome: Codex Alimentarius Commission.

Fallows, S. J. (1988). *Food Legislative System of the UK*. London: Butterworth and Company (Publishers) Ltd.

Food and Agriculture Organization. (1991). Management of Food Control Programmes. In: *Manuals of Food Quality Control*. Rome: Food and Agriculture Organization.

Food and Agriculture Organization/World Health Organization. (1997). *Joint FAO/WHO Consultation on Food Consumption and Exposure Assessment of Chemicals*. Geneva, Switzerland, February, 10–14, 1997. Geneva, Switzerland: World Health Organization.

International Organization for Standardization. (1991). *Development Manual 6, Application of Standards*, 1st ed.

International Organization for Standardization. (1996). International Organization for Standardization. *ISO Bull*.

Lupien, J. R. (1993). Technical harmonization of international trade agreements. *Food Technol* 47, 106–114.

Miller, S. A. (1993). Health, safety, and standards: do we need an international food regulatory institution? *Food Technol* March, 125.

U.S. Office of Technology. (1992). *Global Standards: Building Blocks for the Future*. TCT-512. Washington, DC: U.S. Office of Technology.

Whitehead, A. J. (1994). The need for a strong food control system to ensure international trade. In *Food Science and Technology: Challenges for Africa*. Edited by L. T. Marovatsanga and J. R. N. Taylor.

World Health Organization. (1996). *Food Safety Issues: Guidelines for Strengthening a National Food Safety Programme*. Geneva, Switzerland: World Health Organization.

CHAPTER 10

A View from Consumers

Diane McCrea

INTRODUCTION

International food standards and the work of the Codex Alimentarius are hardly everyday concerns of consumers as they buy their daily foods. Yet Codex Alimentarius and its deliberations in the setting of international food standards do have an impact on ordinary consumers and their daily foods, almost everywhere in the world.

- How does Codex ensure that consumers' interests are protected, that food is safe to eat, and that trade is fair?
- How are consumers' interests represented in the system?
- Does Codex respond to consumers' needs and interests?
- What are the consumers' concerns with the present system?
- Is decision making based on the right principles?
- Is the process too slow to respond to changing needs?
- Is the consumer voice drowned out by the industry lobby?
- Is the process democratic and open?
- Is the system cracking under the strain?

These and other controversial questions will be explored in this chapter.

THE IMPACT OF CODEX

It would be fair to say that most consumers have never given any of the above issues a thought or even heard of the Codex Alimentarius Commission (CAC). Most consumers are unaware of its deliberations; complicated procedures and decisions; and how these impact on the foods they select, prepare, and consume every day. Increasingly, foods are subjected to Codex decisions since these are used as the reference standards and baseline for international food trade rules under the World Trade Organization (WTO) agreements. Foods everywhere will ultimately either directly or indirectly be influenced by Codex, in many different dimensions—safety standards; additives; pesticides; labeling of prepacked foods; and international trade, competition, and pricing—be they foods grown locally or imported from the other side of the world.

The impact of Codex on international trade and consumers' food choices should not be underestimated. This is more pertinent now than when the Joint Food and Agriculture Organization (FAO) and World Health Organization (WHO) Food Standards Conference decided to establish Codex back in 1962. Since December 1993, and the signing of the General Agreement on Tariffs and Trade (GATT), Codex standards have become reference texts used by the WTO for international trade and food safety, and hence, form a basis for international standards.

The Original Remit

The remit of Codex, when first set up by the FAO and WHO, was to "guide and promote the elaboration and establishment of definitions and requirements for foods, to assist in their harmonization and, in doing so, to facilitate international trade" (CAC, 1994). Codex standards also contain "requirements for food aimed at ensuring for the consumer a sound, wholesome food product free from adulteration, correctly labelled and presented" (CAC, 1997a). The balance between the two requirements of Codex, to facilitate trade and ensure consumers are provided with sound, wholesome food, has become increasingly complex and controversial, as will be illustrated.

Prior to the signing of the GATT, little was at stake in Codex in terms of world food trade obligations. Codex decisions were not legally binding on members nor were they formally the rules for international food trade. However, since January 1, 1995, and the Agreement on Sanitary and Phytosanitary Standards (SPS Agreement) under the WTO, changes in the status of Codex texts, standards, guidelines, and recommendations ensued. The SPS Agreement is now binding on all WTO members. Codex standards are officially recognized as reference standards in WTO—whereby should a trade dispute occur, Codex standards would be used to resolve the dispute. As a consequence of WTO and the new status of Codex within it, decision making and the setting of Codex standards have taken on a new importance and become even more controversial.

Public Understanding of Codex

Codex has been subjected to criticism from some consumer groups who argue that the balance of decision making has focused on the facilitation of trade rather than the acknowledgment of other legitimate consumer concerns, including short- and long-term safety issues. Some consumer groups claim that the basis for decision making fails to account for issues of paramount concern for many ordinary consumers the world over. Indeed, it is difficult to explain the complexities of

Codex to the general public, and when this is attempted, there is often incredulity over its dual remit and the Codex emphasis on sound science and narrow interpretation to ensure wholesome products reach consumers worldwide. Wider public concerns about food such as animal welfare and environmental aspects of sustainable food production are beyond the remit of Codex.

A debate about Codex in the broader public arena is long overdue. Codex has done little to extend its reach and make the public aware of its relevance, authority, and significance. A recommendation of the FAO/WHO Conference on Food Standards in 1991 on consumer participation in decision making (para. 33, iii) specifically on information and support suggested that "FAO/WHO consider producing short, accessible summaries of Codex issues and discussions for public distribution." While there is now much greater public access to Codex papers and procedures as published on the Internet, many people find these to be overly complex and often impenetrable. Easy-to-understand summaries on current matters for public distribution would still be welcome. It was also recommended by the CAC in 1999 that FAO, WHO, and national governments work with consumer organizations to improve the dissemination of Codex information to consumers.

Misinformation about Codex, its authority, and procedures has fueled much public mistrust of Codex, particularly in the United States, Europe, and developing countries. This was evident in the media campaign around the Codex Committee on Nutrition and Foods for Special Dietary Uses (CCNFSDU) and its deliberations on vitamin and mineral supplements, when at its meeting in Berlin 1998 it was discussing draft guidelines for vitamin and mineral supplements. These products are marketed widely throughout the world at varying dose levels. There is a real fear that unregulated usage of very high-dose supplements may harm consumers' health. CCNFSDU was beginning to discuss these dose levels and the extent to which restrictions should or should not be introduced for safe upper dose limits for these products.

A German consumer lobby group protesting outside the CCNFSDU meeting had paid for a high-profile media campaign against Codex and (as they thought) its impending restrictive practices on these products. The group appeared to believe that Codex was about to make a decision, which would prevent access to these products and restrict individual consumer choice. However, those familiar with Codex procedures would be well aware that it could be many years before Codex made and implemented any decisions on this matter. Codex uses scientific evidence of safety as the basis for its decision making, and has a remit to ensure that all products on sale are safe and wholesome. In short, there was actually nothing to protest about: Codex is unable, by its very procedures, to act quickly nor is it able to ban products legally on sale in any individual country. Yet Codex does have the responsibility to set safe standards for foods in international trade, and would effectively ban international food trade in any products it deemed unsafe.

Public perception in this case was that Codex was about to act immediately to ban all high-dose vitamin and mineral supplements. Whereas in reality, Codex will continue its laborious discussions on this topic for many more years; it has already been discussing this topic for over five years with very little progress. And the many interpretations of safe upper limits and the debate on the scientific necessity for more than the recommended daily amounts of such nutrients in supplement form are such contentious issues that Codex may never be able to agree on these guidelines. There is a very long debate ahead of CCNFSDU to determine the safe upper limits for vitamin and mineral supplements and to ensure that consumers are protected against any unsafe high-dose products. Meanwhile, the public remains confused.

Consumer Participation at Codex Meetings

Individual consumers are not able to participate in Codex meetings; consumer organizations are, however, eligible to obtain observer status to participate in Codex meetings at the national and international level. Increasing consumer participation has been a specific objective within Codex during the 1990s. How to achieve this has been discussed in many committees, including regional committees, the General Principles Committee, and the CAC.

Consumers International (CI) has represented the broad-based consumer interest in Codex since the early 1970s. At that time, CI was the only recognized international nongovernmental organization (INGO) that represented the broad-based, global, consumer food perspective at a whole range of Codex committees. Other public and special interest groups also participate in Codex as observer INGOs, including Association of European Coeliac Societies, European Heart Network, International Cooperative Alliance, International Baby Food Action Network, and the International Lactation Consultants' Association. More recently, the International Association of Consumer Food Organizations has also become an observer member.

Public interest groups from individual countries are occasionally invited to join their national delegations when attending Codex committees; such as the Community Nutrition Institute in the delegation of the United States or Australian Consumers' Association in the delegation of Australia. In these instances, they, like all other delegates, have to raise their own funds to cover expenses. Recently, Norway, the Netherlands, and India have specifically funded and included a consumer/public interest representative within their country delegations to particular Codex meetings. This is to be commended as an example of supporting the consumer interest and actually facilitating consumer participation since it is the financial resources that are often the greatest limiting factor to increasing consumer participation in Codex.

CI (previously known as the International Organization of Consumer Unions [IOCU]) has represented the consumer interest in Codex for many years. Basic consumer principles are used for CI's work, developing its policy with members, in all parts of the world on all consumer issues, not just on food. President John F. Kennedy first elaborated five key consumer principles; these have been modified over time and form the basis of the fundamental rights of consumers used by CI in its policy work. These are now:

1. The right to satisfaction of basic needs
2. The right to a healthy environment
3. The right to safety

4. The right to redress
5. The right to be heard
6. The right to be informed
7. The right to consumer education

CI, as an umbrella organization, has over 220 member consumer organizations in over 100 countries worldwide. It focuses and campaigns on a broad consumer agenda, including trade and economics, environment, health, technical standards, and food. Resources are limited, however, and must be prioritized within the overall food arena and the Codex agenda.

At the international level, CI has to prioritize its Codex work; it participates in general subject committees on horizontal subjects such as food labeling, general principles, pesticide residues, food additives and contaminants, and food hygiene. CI increasingly focuses on regional coordinating committees. It is impossible for CI to fund and send delegates to all the Codex committees; resources unfortunately do not permit this level of involvement.

Few delegates from consumer and public interest groups, as well as some member country delegations, particularly from developing countries, are able to afford to participate fully in Codex due to lack of both financial and human resources. Preparing for, attending, and following up on Codex meetings demand considerable time, commitment, and financial resources from all delegates, be they government officials, public interest, or industry representatives. CI has long campaigned for a mechanism for funding consumer participants, at both the international and national level. This is a serious issue that impedes further participation, particularly among fledgling consumer groups in developing countries.

At the CAC in 1997 the commission agreed to investigate the establishment of trust funds to facilitate greater participation from developing countries and consumer representatives, but these funds have not been forthcoming for consumer participation. Indeed, subsequently at the CAC in 1999 it was noted that there was little support to identify funds for consumer INGOs in Codex; it was agreed that any resources available should be directed first to developing member countries of the CAC.

Consumer Participation at National Codex Consultative Committees

At the national level, members of CI and other public interest organizations are often involved in national Codex consultative committees (NCCCs). These NCCCs were set up as a priority by Codex with the aim of coordinating views from the whole range of organizations and interests from the public and trade sectors, at the national level. Many countries have established such committees (e.g., the United States, South Africa, United Kingdom, Australia, and India). Many, especially those in developing countries, have not yet done so.

Even where NCCCs exist, participation can still be a problem for consumer and public interest groups. Resources are needed to attend meetings, follow the issues, and be able to contribute to the debates. There are particular problems in developing countries with less well-established consumer organizations that often need training before they can effectively participate and contribute to the process. CI set up a three-year training program in 1999 to train member organizations in developing countries about Codex and how to participate effectively within it.

Codex needs to galvanize all its members to set up NCCCs as a matter of urgency, and to improve contacts with consumer and public interest groups at the national level. Systems need to be set up by governments to provide a dialogue; to involve a wide range of interest groups; to explain the national positions, including how these are agreed upon; and to give feedback on Codex meetings and progress on items of particular national interest.

Even where NCCCs have been set up, there are inconsistencies between countries; some have not allowed consumer and public interest groups to participate, while others actively encourage this. CI, in coordinating with its members in many different countries, finds inconsistent procedures that cause concern, confusion, and barriers at the national level. This was evident when CI surveyed its members in May 1999; only 46% of respondents reported that regular consultations took place with their national governments.

National Codex contact points should be proactive and encourage a wider involvement of consumer and public interest groups in its activities—disseminating information, soliciting alternative opinions, and facilitating education and training on Codex and its relevance to national and international food standards. At the moment, some of these contact points merely operate as a distribution point for Codex documents. Governments need to extend their role to fulfill this commitment and obligation for their consumers and traders alike. At the CAC in 1999, it was agreed to consider the development of a checklist of measurable objectives to assess consumer participation in Codex at the national and international levels; this would be taken up for further review by the Codex Committee on General Principles (CCGP).

MEMBERSHIP OF CODEX

When Codex was set up in 1963, there were 38 original members; in 1999 membership had risen to 165 member countries, representing 98 percent of the global population. In addition, there were 125 observer INGOs in 1999.

Membership of Codex has increased steadily since the signing of GATT and the setting up of the WTO. As more food is traded internationally between more countries, so Codex membership continues to rise. The most recent country to join in 1999 was Namibia. The original members were mainly from developed economies, but gradually and, more particularly, since the signing of GATT, membership has steadily increased from developing countries.

Membership in WTO means that member governments should use Codex for their reference standards in international trade. But it does not necessarily mean that they are actively required to participate in the Codex decision-making process. At the CAC in Rome, June 1999, 103 of 165 eligible member countries attended. All government delegations have full membership, authority to participate in decision making, and equal voting rights, which are applicable for all member countries, large or small. This raises serious concerns from those delegations, mainly from developing countries, lacking the resources and expertise to fully participate in Codex, even though they are equal members with other more affluent countries and industry observers with more resources to devote to Codex.

At the meeting of CCNFSDU, in Berlin in 1998, developing countries were in the minority with very few delegations from the Africa and Asian regions. A major discussion took place on the standards for infant formula products. The problems experienced by consumers in developing countries when using these products are very different indeed from those experienced in the developed world. The consumer voice from developing countries was underrepresented and overwhelmed by the much larger industry participation in this meeting.

Other nongovernmental interests can be represented by observers, providing they are from recognized INGOs. These include food industries and businesses, professional bodies, specific interest bodies, nonmember governments, and public interest, nongovernmental organizations. The range extends from the Confederation of International Soft Drinks Association to the International Dairy Federation, to the Association of European Coeliac Societies, to the International Organization for Standardization, right through to the European Commission and CI.

INGO observers represent the whole range of interests at Codex, but the majority of observers are industry-funded. When the matter of INGO participation was discussed at the CAC in 1997, there were at that time 111 recognized bodies approved to participate as INGO observers; over 100 of these are industry-funded. Of the 34 listed INGO observers who attended and were listed in the participants list of the CAC in Geneva, June 1997, 26 were industry-funded, 3 were professional organizations, and 5 were public or special interest INGOs.

A detailed analysis of representation and funding within Codex committees was carried out by Avery et al. (1993) in *Cracking the Codex*. This report analyzed participation of observers in Codex proceedings and concluded that many reforms were necessary to redress the balance of power between industry and public interest groups. Indeed, at the CAC meeting in 1997, when discussing the involvement of INGOs in the work of Codex, the delegations of the United Kingdom and Norway stated that "the role of consumers was different from that of trade bodies. In view of the changing status of Codex documents, it was important to avoid any perception of Codex being unduly influenced by commercial interests" (CAC, 1997b). In addition, the definition of INGOs needs to be considered further to differentiate between public interest and commercial interests, so that the balance of representation can be more evenly applied within Codex.

PRINCIPLES FOR DECISION MAKING

Difficult issues have confounded decision making in the CAC, particularly since the WTO was established. Different perspectives and interpretation of the principles for making decisions, be they based on science, safety, trade, or consumer concerns, have caused much debate and thwarted decision making on several occasions.

Consideration of the maximum residue limits (MRLs) for growth-promoting hormones provoked one of the most controversial and extended debates in the history of Codex decision making. And however contentious the final decision on this matter (MRLs were finally adopted by a secret ballot at the CAC, in Rome in 1995), many issues of principle for decision making were resolved en route (Chapter 8).

Wirth (1994) contends that the long-running disagreement between the United States and the European Union, over growth-promoting hormones in meat, actually prompted the development of the SPS Agreement in the Uruguay Round. This specifies that "sanitary and phytosanitary measures must be based on scientific principles and . . . not maintained without sufficient scientific evidence" (SPS, 1994).

Conflict erupted in 1991 at the 19th Session of the Commission with the decision not to adopt MRLs for growth-promoting hormones. Subsequently the 20th Session of the Commission in 1993 decided to delay decision making again until the status of science in Codex policies and procedures was clarified (CAC, 1993). Codex's inability to move forward and make a decision on growth-promoting hormones prompted the clarification of its decision-making principles.

The preeminence of scientific principles (which underscores the SPS Agreement) and recognition that other factors need to be taken into account in relation to Codex decision making were elaborated and finally agreed at the 21st Session of the Commission in Rome in 1995. A lengthy and controversial debate eventually confirmed that Codex standards and other texts should be based on the principle of sound science, culminating in the adoption of four statements of principle for the guidance of Codex work (CAC, 1995b). These four statements are shown in Exhibit 10–1.

Adoption of the statement of principle for decision making was highly controversial. Spain, on behalf of the member countries of the European Community, expressed opposition to the CAC's decision on the adoption of this statement (CAC,

---

Exhibit 10–1 Statement of Principle Concerning the Role of Science in the Codex Decision-making Process and the Extent to Which Other Factors Are Taken into Account

1. The food standards, guidelines, and other recommendations of Codex Alimentarius shall be based on the principle of sound scientific analysis and evidence, involving a thorough review of all relevant information, in order that the standards ensure the quality and safety of the food supply.
2. When elaborating and deciding upon food standards, Codex Alimentarius will have regard, where appropriate, to other legitimate factors relevant for the health protection of consumers and for the promotion of fair practices in food trade.
3. In this regard it is noted that food labeling plays an important role in furthering both of these objectives.
4. When the situation arises that members of Codex agree on the necessary level of protection of public health but hold differing views about other considerations, members may abstain from acceptance of the relevant standard without necessarily preventing the decision by Codex.

Source: Adapted with permission from *Report of the Codex Alimentarius Commission*, 21st Session, Rome, © 1995, Food and Agriculture Organization.

1995b). CI also objected to the narrow interpretation and preeminence of science in the decision-making process, arguing that science cannot be regarded as absolute, especially when considering human health and the establishment of standards for additives, pesticides, and residues, where long-term risks cannot be fully assessed.

Sound Science

Irrefutable within Codex is that decision making is based primarily on sound scientific analysis and evidence. But what is sound science and who decides? This question in itself raises fundamental consumer concerns. What if science does not have all the answers, or there is inadequate analysis or conflicting evidence, or the science is biased or perceived to be so? Good science costs and takes a serious long-term investment. Increasingly, science is being funded by industry instead of by governments, which raises serious doubts over its impartiality.

Sheila McKechnie, director of Consumers' Association UK, speaking in the Caroline Walker Lecture 1998 stated, "If research doesn't give us a clear picture—we should delay making a decision until the research can provide us with a clearer picture. In effect this means applying the precautionary principle. But if in any risk decision-making process the level of proof required for action is causal and verifiable then this itself acts as a barrier to the implementation of the precautionary principle" (McKechnie, 1998).

Debates on safety have focused on scientific consensus and the nature of sound scientific evidence. Recent history, particularly the safety of beef in relation to bovine spongiform encephalopathy (BSE) in the United Kingdom, has challenged some of these assumptions to their utmost. The research agenda has been unable to provide the definitive answers quickly enough and reliable quantification of long-term risks has thus far proved impossible. Decision making based on sound science alone is unreliable in such a context, and needs to be adapted to consider broader issues and risk assessment.

Codex decision making on setting limits, for example on pesticide residues or food additives, has in the past focused on safety assessment of the pesticide or additive in isolation. More sophisticated paradigms, addressing issues of multiple exposures to multiple pesticides, differential intakes for children, variability in susceptibility of some consumers, short-term as well as long-term effects, and other issues need to be addressed. CI has called upon Codex, particularly the Codex Committee on Pesticide Residues, to improve its risk assessment, especially to develop explicit risk assessment policies that clearly identify components of the process where uncertainties exist and where assumptions and value judgments must be made, and has called for guidance on how to apply these. The CAC has decided to base health and safety aspects of Codex decisions on risk assessment. How risk assessment is defined and implemented at the international and national level and Codex standards reviewed to take account of this are major challenges.

Other Legitimate Factors

While examining and establishing specific guidelines for decision making, factors in addition to science need to be considered and defined. When first attempting to clarify and define them, the CAC originally noted in discussions in 1993 that "these factors included legitimate consumer concerns, animal welfare, fraudulent or unfair trading practices, labeling and other ethical and cultural considerations while stressing the pre-eminence of science in Codex procedures" (CAC, 1993).

However, final agreement on the statement of principle for decision making narrowed these factors to point 2 in Exhibit 10–1 focusing "where appropriate, to other legitimate factors relevant for the health protection of consumers and for the promotion of fair practices in food trade."

Further clarification and the application of these other legitimate factors have been called into question, notably over the approval of MRLs for the hormone bovine somatotropin (BST). Lack of clarity impedes and obscures decision making and fosters ambiguous interpretation of trade and economic factors as well as undermines consumer concerns and confidence in the process and its outcomes.

Clarification of the other legitimate factors is critical, controversial, and long overdue. Some delegates at the CCGP in Paris, 1998, called for a revised remit of Codex near to that originally proposed in 1993, to take on board the widest possible interpretation. As part of those discussions, CI took a more pragmatic approach within the current framework, accepting that efforts to reform the current remit need to be pursued elsewhere.

In its very detailed papers presented for discussion at the CCGP in 1998, updated for CCGP in 1999 (CI, 1998), CI projected that other factors are inescapably part of Codex decisions and include a variety of subjective value judgments and social choices in the application of risk analysis. Among the most important are how to treat scientific uncertainty, and perceptions as to which risks are significant. Others include economic concerns, such as the feasibility of risk management options, and the benefits of the activity or substance that poses the risks. Ethical issues, such as the rights and responsibilities of all the parties involved in the risk management process, also enter the picture.

In Codex decisions with respect to food labeling, CI proposed that other factors apply. In particular, consumers' need and expressed desires for information on a food issue must be acknowledged and given weight. Consumer preference per se is not a basis for Codex to prohibit or limit the use of a food substance or technology. But recognition that consumers do have preferences, based on many factors that cannot be considered by Codex, provides a basis for labeling so that consumers can exercise their preferences in the marketplace and manage their own risks. This is particularly so with the controversial new technology of genetic modification of foods. CI has called for and campaigned within Codex for mandatory labeling to ensure consumers are provided with the necessary information to make informed food choices.

With respect to the specific case of BST, CI presented detailed analysis, in Annex 1 of its paper (CI, 1998), of some factors other than science that it believes are almost certainly part of the Codex risk analysis process on BST. Such aspects include how to deal with the major uncertainties in the scientific data on BST risks and judgments of whether the risks are "significant," being highly subjective and likely to be based on other factors, as well as on scientific evidence. Many other factors that appeared likely to be part of decisions at some point in the Codex risk analysis on BST were identified, and ways to clarify those factors were suggested. In addition, the CI paper presented a determination of whether these other factors have played a transparent and appropriate role in the process.

CI has argued that, particularly with regard to BST, consumers are opposed to its approval since it has no benefit for them; it would not improve milk quality, might lower food safety, and create animal welfare problems; and that the importance of taking legitimate factors other than sound science and evidence into consideration is also important (CAC, 1997a). At the CAC, it was recognized that there was in fact no consensus for the approval of BST; its safety and use are therefore determined at the national level and ranges from full approval in for example the United States to an absolute ban on its use in the European Community.

In relation to Codex standards containing "requirements for food aimed at ensuring the consumer a sound, wholesome product free from adulteration, correctly labeled and presented" (CAC, 1995a), CI believes that any subsequent Codex approval for the use of BST without adequate labeling of the products would be a negation of this basic Codex requirement. Likewise, in April 1997, several delegations at the Codex Committee on Food Labeling when discussing biotechnology were of the opinion that the food safety approach did not address the concerns of consumers in areas such as ethics and environmental protection. The delegation of Norway expressed the view that "the issues associated with modern biotechnology went beyond information about product characteristics, that the rights of consumers to make their choice should be respected even if this meant broadening the basis for labeling requirements, and that labeling was the only means to ensure consumer confidence in this area" (CAC, 1997c).

The Precautionary Principle

In his opening statement at the October 1999 FAO Conference on International Food Trade Beyond 2000, Melbourne, Australia, Mr. Hartig de Haen, assistant director general of the FAO, noted that "when dealing with food contamination, whether from environmental sources or of microbiological origin, scientific data and evidence of safety are often absent or incomplete, and prudence dictates that adequate precautionary measures must be taken."

A precautionary approach counsels decision makers to err on the side of public health protection when formulating standards in cases where there is significant scientific uncertainty, or where the potential public health consequences of making an error are significant. This is a prudent approach and rests on the ethical principle that it is "better to be safe than sorry" when public health is at stake.

In terms of both health hazards and economic impacts, the cost of not taking action because of uncertainty may be far greater, in the long run, than the costs of prudent precautionary action. The case of BSE (or Mad Cow Disease) in the United Kingdom is one recent example, in the food safety context, of an instance in which early preventive measures based on the precautionary principle may well have averted extraordinary social and economic costs to society.

As a practical matter, the most critical question with respect to the precautionary principle is probably not whether it is appropriate to consider it, but rather when it is appropriate to invoke it. CI agrees that a precautionary approach is justified within Codex when there is uncertainty as to the exact nature and magnitude of the risk. The Secretariat, in paper CX/GP 98/10 (*Review of the Statements of Principle and the Role of Science and the Extent to Which Other Factors Should Be Taken into Account*), offers some criteria that might determine when to apply the precautionary principle, including having an incomplete risk assessment and difficulty in qualifying the risk or in identifying appropriate risk management options. The paper states that, "Although there is no generally recognized definition, the precautionary principle is supposed to apply in cases where the scientific evidence is not conclusive enough to determine a level of protection, but there is necessity to apply measures for the purpose of protecting the public health." The paper goes on to state that, "In any case the precautionary principle is related to the health risk and is intended to assess uncertainty or incomplete scientific evidence, which cannot apply in the case of BST as the scientific basis clearly exists." This implies that the application of a precautionary approach, with reference to BST, is ruled out since it has been scientifically agreed that no health problems have been identified—this is disputed by some. This is not the same as proving that BST is safe. The matter of defining the precautionary principle is to be discussed further in the CCGP.

All risk assessments and risk management strategies contain scientific uncertainty, some more than others. In some cases, not enough may be known even to properly tell what the critical uncertainties are. How much uncertainty is acceptable in any given case is a social value judgment, and so nonscientific factors are considered in making that decision in the risk management process. A decision by risk managers that the degree of uncertainty is too large, given the potential consequences of an error, would be a sound basis for invoking the precautionary principle.

Regardless of whether Codex accepts the precautionary principle as a legitimate factor in decisions (which CI believes it should), CI believes that it is necessary to make greater efforts to ensure that the scientific uncertainties in risk assessments are transparent to risk managers. In the Codex context, for example, expert advisory bodies might be asked to provide more explicit assessments of the uncertainties inherent in the scientific data they have reviewed and the analytical approaches they have chosen, and to indicate the impact of the uncertainties on their decisions. For example, expert bodies should be asked to identify and quantify key uncertainties regarding risks, assess the degree of confidence they have in their key scientific conclusions, justify their reasons for rejecting alternative scientific evidence, and explain the possible public health consequences if any of those conclusions later turn out to be incorrect.

Risk Assessment

Codex standards aim at protecting consumers' health, but assessing risks to health is complex and cannot be based on known scientific facts alone. Incorporating risk assessment principles into Codex decision making is a major challenge facing all Codex committees. Codex in its statement of principles relating to the role of food safety risk assessment states (CAC, 1997c):

1. Health and safety aspects of Codex decisions and recommendations should be based on risk assessment, as appropriate to the circumstances.
2. Food safety risk assessment should be based soundly on science, should incorporate the four steps of the risk assessment process, and should be documented in a transparent manner.

Deciding upon the inherent risks (e.g., in new food-processing techniques such as genetic modification, or in older techniques such as pasteurization of milk, or risks associated with levels of contamination in natural mineral water) has all been extremely controversial in Codex. All these issues raised fundamental differences in philosophy and acceptance of the level of risks. Cultural acceptances of risks differ fundamentally at the national level. To take but one example: the French accept the risks associated with eating unpasteurized cheeses and dairy products, while the United States considers such risks unacceptable and banned the sale of all such products in interstate trade since 1987 to provide adequate consumer protection. Consumer groups in the United States have lobbied for a tough stance on this issue in Codex. The Center for Science in the Public Interest, a North American-based INGO observer in Codex, stated that "U.S. negotiators must toughen their stance . . . or face the possibility that U.S. consumers will be subjected to unsafe imported food products ranging from unpasteurized cheese to poorly inspected meat" (CSPI, 1997).

There is no compromise in the different interpretations and acceptance of inherently the same risk to health. Where fundamental differences of opinion occur, Codex procedures must ensure that decisions are based on sound science and risk assessment, that decisions are not being influenced by trade interests or protectionist measures, and that all consumers are adequately protected. Consumers expect their foods to be safe and to trust national and international standards' bodies such as Codex to set the standards to protect them from unsafe foods. Where Codex standards are not accepted, bilateral trade agreements can be negotiated.

Expert Technical Opinion

Codex depends on expert technical bodies, convened by FAO and WHO to advise on food additives and contaminants, veterinary drugs, and pesticide residues. These are respectively the Joint FAO/WHO Expert Committee of Food Additives (JECFA) and the Joint Expert Meeting on Pesticide Residues (JMPR); both provide technical scientific reports of risk analysis for Codex to base its decisions upon.

The competence of these committees and their processes to deal with all the issues on which Codex seeks expert opinion has recently been questioned. Are the experts technically competent in all the relevant areas? Are they free from commercial bias, acting independently and impartially? CI, in its response to the Codex Review (Codex Circular Letter, 1998/2GP) of the Code of Ethics for International Trade, called for the criteria for the selection of experts to be made publicly available. In addition, CI has many times called for the declarations of interests of all selected experts chosen to advise on expert and/or technical advisory groups and panels to be made publicly available. This does not necessarily mean to exclude those who declare an interest, merely to make this fact publicly known.

Both committees are closed to observers since in the view of FAO and WHO "the presence of observers at these meetings could establish an atmosphere of influence" (CAC, 1997d). This together with the secrecy of decision making and the lack of public information on members' interests has brought the credibility of these committees and their impartiality into question by public interest groups. Avery and others (1993) recommended that, "In order to maintain credibility in the assessments of these committees, it is essential that the FAO and WHO draw up and publish, in consultation with public interest organizations, a new set of criteria for selecting Expert Committees and other groups. . . . Members of Expert Committees should be required to declare in a public register every meeting and any funds received directly or indirectly from industry." This has not yet happened, nor is it any nearer to happening than it was in 1993.

Reference of an issue to expert opinion is requested often, but is also denied. Both JECFA and JMPR have full agendas and are unable to take on all issues that some Codex committees would like to request. However, recently, additional special conferences have been held to discuss matters pertinent to Codex. For example, the United Kingdom hosted an international conference in November 1998 to discuss worldwide controls on the variability of pesticide residues and acute dietary risks. The conference sought to achieve a global understanding of the implications of variable residue levels and promote development in pesticide controls to take account of variability. These deliberations will feed into the Codex process and help to focus discussion and decision making among experts, over and beyond what is possible within Codex procedures. Deliberations were limited to those invited and those able to attend, which naturally raises some questions over accessibility to decision making. Any recommendations from such select forums will need to be fully scrutinized and adopted by a formally constituted Codex committee.

Consensus

Decisions in Codex are normally made by consensus. Indeed, there is a specific reference in the procedural manual (CAC, 1995a) that chairpersons "should not ask the committee to proceed by voting if agreement on the Committee's decision can be secured by consensus." However, the meaning of the term *consensus* is not defined in any Codex procedures, and it has recently proven difficult to achieve in practice.

At the 22nd Session of the CAC, in Geneva in 1997, consensus could not be reached, and two major decisions were made by roll call votes of the members present. It was even suggested that since no consensus could be reached, the Revised Draft Standard for Natural Mineral Water should be decided by secret ballot. However, a roll call vote was taken, and a slim majority adopted the standard. The tally of votes showed 33 in favor of adoption, 31 against, and 10 abstentions. Thus a vote by 33 members was binding on all Codex members—161 at that time. This decision provoked huge controversy: a process of adopting Codex standards whereby a mere 2 vote majority effectively accepts this as a worldwide standard was challenged as undemocratic for all those members not present and unable to vote. The delegation of the United States stated that it was regrettable that the decision had been made by a vote and that "the standard ignores public health protection . . . and creates a barrier to international trade" (CAC, 1997c). Several delegations expressed concern that a decision had been made by voting and "stressed that the Commission should try by all appropriate means to attempt to take such important decisions on the basis of consensus" (CAC, 1997c).

Codex issues are often fiercely debated in an attempt to reach consensus. However, Codex procedures also allow for decisions to be made by voting and even by a secret ballot. At all sessions of the CAC since 1991, consensus at some time or another has proved illusive. Some decisions of major importance have been taken on the basis of a vote. So contentious was the decision making over the adoption of draft MRLs for five growth-promoting hormones at the 21st Session of the CAC, in Rome in 1995, that the facility to use a secret ballot (at the request of the delegation of the United States) was evoked. The secret ballot procedure was, in the view of the observer of the European Community, regrettable and contradictory to the CAC's decision to increase transparency. CI stated that while frequent mention was made to transparency, openness and the necessity for consumer confidence in Codex procedures, neither of these things had been improved in the decisions reached nor in the way in which some of the decisions were made at this session of the CAC (CAC, 1995b).

How consensus is to be achieved in an open, transparent manner when there are such fundamental differences of opinion is difficult to imagine. This is being discussed by the CCGP and presents a major challenge for Codex to overcome and progress in the development of all future standards. Decisions, which are binding on all Codex members, can, as shown above, be taken by a small majority of the members present. Voting by simple majority is increasingly being criticized, within and beyond Codex's membership. Any WTO member opposed to Codex standards does of course have the option of setting higher standards, but if challenged on such a decision could also risk invoking the WTO disputes mechanism.

Developing countries are very active in the debate to improve their participation in the decision-making process. India has argued that a fundamental rethinking is needed to ensure that the views of developing countries, especially those that cannot afford to attend Codex committees, should be incorporated into the decision-making process, perhaps even by allowing postal voting. While there are many practical considerations to consider, these matters are being given due consideration in the CCGP's review of measures to facilitate consensus, which is urgently needed.

Unequal Voices

Consumer and public interest groups have no voting rights as observers at Codex; INGOs are not full parties in the decision-making process. They are, however, able to present written comments and are allowed to address Codex committees.

Protocol does not allow interventions from INGOs at the appropriate point in a debate, but rather after all other member delegates have spoken on an issue. This means that consumer and public interest groups (like all other INGOs) are at a significant disadvantage in being unable to participate at the appropriate point in the discussions. INGO interventions are restricted to tail end comments or observations after all member countries have made their interventions. Even when the agenda item being discussed was improving consumer participation in Codex procedures, on some occasions comments were taken from all member countries before CI. While INGOs do not expect to participate in decision making directly, they do expect to be equal partners in the discussions, to have an equal voice, and not be relegated to the last word just before discussions are closed. This outdated protocol should be amended: INGOs should be able to participate fully in all discussions, just as, at the discretion of the chair, any other member can.

Good Governance

Given the responsibilities of Codex under WTO and the new public interest in these affairs, Codex should operate and be seen to operate under the tenets of good governance and incorporate principles of openness, transparency, and accountability into all its procedures and practices. While considerable progress has recently been made in this area, nevertheless, the principles of openness and transparency in all decision making and procedures should be reviewed and yet further revised. Consumers now expect and demand broader democratic accountability from decision and policy makers. Indeed this philosophy was included in a recommendation adopted at the FAO Conference on International Food Trade Beyond 2000. This recommendation stated "that policies should be adopted wholly consistent with the need for an independent and transparent risk assessment process in particular in relation to the selection of experts, the working procedures and the tightening of the conflict of interest requirements" (FAO, 1999). CI believes that a declaration of interests of all experts should be placed in the public arena; this would ensure that any potential or perceived conflict of interest can be addressed in the open and not behind closed doors.

INGOs are not allowed to participate as observers in the Executive Committee or technical and expert committees of Codex and FAO/WHO. CI has long campaigned for these committees and their deliberations—made in the public interest—to be made accessible to public interest groups. There has in the past been great resistance to this and, as a result of this secrecy, mistrust in Codex and its procedures has been fueled.

The CAC meeting in 1999 noted that the three types of INGOs—consumer and public interest groups; the food industry, trade, and marketing organizations; and professional and scientific international nongovernmental organizations—have made valuable contributions to the work of Codex and that to enhance the transparency and credibility of Codex decisions, consideration could be given to involving these groups in the work of the Executive Committee. No firm agreement was made to approve this at the CAC in 1999, but it was agreed to ask the CCGP to develop proposals along these lines for further consideration by the CAC at a later date.

Access by consumer observers may still be denied in some committees, but public access to Codex documents has improved greatly since Codex has developed its Internet provision, which has been used to publish and distribute its papers. This is a major step forward for all those with access to the Internet. Further use of the Internet will facilitate greater public access to its papers, proposals, and decisions. However, how useful and accessible these proceedings are in their current format, for those uninitiated in Codex and its jargon, is debatable. More effort should be made to make available easy-to-digest summaries of Codex work and its decisions.

Not all member countries and observers have access to the Internet, and, therefore, conventional means of distributing papers will still be necessary in the future. In addition, papers that are supposed to become available before CAC meetings are frequently late being released. Whether due to late presentation of a position by a member country or late preparation of a paper by the Secretariat, this persistent problem limits effective consideration of the papers by delegates. Inadequate time to prepare a detailed response to important items is a restriction on effective participation and progress. Some very important papers have only been available for few weeks before some committees meet, thereby making it impossible for some delegations to consult with their constituencies and prepare detailed responses. Extreme pressure of work within the small Codex Secretariat is a factor. In addition, many of the discussion papers are initially prepared by member governments, which are also under pressure. The issues are complex, and preparing a discussion paper can be a major responsibility not to be undertaken lightly. However, late distribution of papers frustrates the process, the delegates, and the outcomes.

Codex relies upon member countries to host Codex meetings by providing support for the hosted Committee, including the local secretariat, the venue, and the translation services. Hosting a Codex committee can be a major expense for a government. The resource implications have become apparent, for example, at the CCNFSDU hosted by Germany. The agenda is long, and progress is very slow, yet there is reluctance to hold meetings more often due to resource implications. Currently, CCNFSDU meets once in a two-year Codex session, and requests for a meeting every year have not been successful. Even providing adequate translations in Spanish, an official Codex language, was a problem at the 1998 session, and thereby excluded effective participation from Spanish-speaking delegations.

Codex procedures are lengthy; decisions take many years before final agreement. When committees meet only once in a two-year Codex session, any delay in publishing discussion papers is frustrating, but can also unduly prolong the process.

The laborious nature of Codex procedures is often criticized; this has been particularly so with regard to developing guidelines for the labeling of genetically modified foods. These discussions were initiated back in 1993, but are still at the initial steps of Codex procedures in the food labeling committee. The agreement and adoption of any such guidelines are likely to take many more years. Meanwhile genetically modified foods are being produced and entering the global food chain, without any requirement for labeling under Codex rules. Codex with its slow mechanism is unable to respond to rapid developments in the marketplace and/or scientific advances; its most rapid systems take years to enact.

At the CAC 1999, in order to deal with some urgent matters more expediently, it was decided to set up new procedures for ad hoc intergovernmental task forces. These task forces are charged with developing standards and guidelines on specific topics within the shorter time frame of up to four years. Two of these task forces of particular interest to CI—Foods Derived from Biotechnology and Good Animal Feeding—were set up in 1999 to address these matters and present recommendations to the CAC in 2003. The resource implications for host countries, delegations wishing to participate in yet another committee, and the Secretariat cannot be underestimated, but nevertheless these priority matters need to be dealt with. In some cases, where there is a limitation on resources for Codex work, and particularly for CI, this will mean a reexamination of future priorities, resource allocation, and participation in Codex work.

CONCLUSION

Codex is gradually adapting and facing up to the challenge of its new role and responsibilities under WTO. Increasing effective consumer participation at the international and national level is still a major challenge. More consumer groups are becoming aware of the importance of Codex and are seeking INGO status; their participation is to be encouraged. Redressing the imbalance, perceived and real, between trade and public interests and ensuring safe food for consumers is the real challenge for the future. Consumers need to have confidence that their concerns are paramount and not overruled by commercial interests. Only when the procedures, remit, and criteria for decision making put consumer concerns at the forefront, taking into account short- and long-term risks, the precautionary principle and its overall application in risk assessment, the other legitimate factors in decision making, and extending the remit of Codex to acknowledge broader consumer concerns will there be improved consumer confidence in Codex. Codex is gradually moving in the right direction. For some, this progress is far too limited and far too slow to meet the growing consumer interest in food safety and quality, international food trade, and standards, especially following immense public interest in the Seattle WTO Ministerial Conference. The challenges have been clearly identified by consumer and public interests for Codex to conquer and champion in the next decade.

REFERENCES

Agreement on the Application of Sanitary and Phytosanitary Measures. (1994). *Agreement on the Application of Sanitary and Phytosanitary Measures*. Uruguay Round of Multilateral Trade Negotiations. 33 I.L.M. 9 Final Act, para. 6. Geneva, Switzerland: World Trade Organization.

Avery, N. et al. (1993). *Cracking the Codex*. London: National Food Alliance.

Center for Science in the Public Interest. (1997). News release on the publication of its report, *International Harmonization of Food Safety and Labelling Standards*. June 17, 1997. Washington, D.C.

Codex Alimentarius Commission. (1993). *Report of the Codex Alimentarius Commission,* 20th Session, Geneva. Rome: Food and Agriculture Organization.

Codex Alimentarius Commission. (1994). *This Is Codex Alimentarius,* 2nd ed. Joint FAO/WHO Food Standards Program. Rome: Food and Agriculture Organization.

Codex Alimentarius Commission. (1995a). *Procedural Manual,* 9th ed. Joint FAO/WHO Food Standards Program. Rome: Food and Agriculture Organization.

Codex Alimentarius Commission. (1995b). *Report of the Codex Alimentarius Commission,* 21st Session. Rome: Food and Agriculture Organization.

Codex Alimentarius Commission. (1997a). *Procedural Manual,* 10th ed. Rome: Food and Agriculture Organization.

Codex Alimentarius Commission. (1997b). *Report of the Codex Alimentarius Commission,* 22nd Session. Rome: Food and Agriculture Organization.

Codex Alimentarius Commission. (1997c). *Report of the Codex Committee on Food Labelling.* ALINORM 97/22a. Rome: Food and Agriculture Organization.

Codex Alimentarius Commission. (1997d). ALINORM 97/8, Agenda item 9, para. 15. Rome: Food and Agriculture Organization.

Consumers International. (1998). *The Role of Science and "Other Factors" in Codex Decisions.* Discussion paper prepared for the Codex Committee on General Principles, Paris 1998 and London 1998, updated 1999.

Food and Agriculture Organization. (1999). *Report of the Conference on International Food Trade Beyond 2000: Science-Based Decisions, Harmonization, Equivalence and Mutual Recognition.* Melbourne, Australia. Rome: Food and Agriculture Organization.

McKechnie, S. (1998). *The Nanny State.* The Caroline Walker Lecture 1998. London: The Caroline Walker Trust.

Wirth, D. A. (1994). *The Role of Science in the Uruguay Round and NAFTA Trade Disciplines.* Environment and Trade Series. Geneva, Switzerland: United Nations Environment Program.

PART IV

Codex and Other International Standards

CHAPTER 11

The Codex General Standard for Food Additives—A Work in Progress

Dennis Keefe, Paul Kuznesof, Susan Carberry, and Alan Rulis

SCOPE AND PURPOSE OF THE GENERAL STANDARD ON FOOD ADDITIVES

The Codex Alimentarius Commission (CAC) is developing an international standard for the use of food additives in foods generally. The Codex Committee on Food Additives and Contaminants (CCFAC) is the forum in which the standard is being developed. When fully elaborated, the Codex General Standard for Food Additives (GSFA) is intended to set out the conditions under which food additives may be used in internationally traded foods. Thus the development of the GSFA is one of the most important standard-setting activities within Codex and represents a significant historical shift in Codex's focus from commodity standards to general subject standards. The Agreement on the Application of Sanitary and Phytosanitary Measures (SPS Agreement) of the General Agreement on Tariffs and Trade (GATT) recognizes CAC as the international standard-setting organization in the area of food. As a result, the GSFA will be recognized as an international standard under the World Trade Organization (WTO) and the use of food additives in accordance with the GSFA will enjoy a presumption of safety by the WTO.

The Netherlands hosts and chairs the CCFAC. Traditionally, the CCFAC has met annually, usually in March, in The Hague. In 1991, the CCFAC established an ad hoc GSFA Working Group (GSFA WG) with the United States as chair. Since 1991, the CCFAC has reestablished this ad hoc working group with the United States as chair. Since its inception, the GSFA WG has convened just prior to the plenary session of the CCFAC.

This chapter provides a summary of the current status of the GSFA, the milestones in its development, and the tasks that remain for full elaboration as a Codex standard. In this context, the GSFA is a dynamic entity that will require continual revision to keep it current with scientific knowledge, to ensure that it protects the health of consumers, and to ensure that it does not unnecessarily restrict the international trade of safe foods.

Source: Reprinted from the United States Food and Drug Administration.

VERTICAL AND HORIZONTAL STANDARDS

The Codex Alimentarius (Codex) was organized in 1962 under the direction of the Joint Food and Agriculture Organization/World Health Organization (FAO/WHO) Food Standards Programme. Its mission is to protect the health of consumers and ensure fair practices in the food trade. One means by which Codex carries out its mission is to develop food standards through an eight-step[1] elaboration process (CAC, 1997a).

Prior to 1991, the standard-setting activity of Codex focused on the development of commodity or so-called vertical standards (e.g., Canned Plums [Codex Standard 59–1981] [CAC, 1995j] and Wheat Flour [Codex Standard 152–1985] [CAC, 1995i]). Codex commodity standards typically have provisions for labeling, the use of food additives, maximum residue levels for pesticides and veterinary drugs, and maximum levels for contaminants. In the development of vertical standards, commodity committees (e.g., the Codex Committee on Processed Fruits and Vegetables) defer to the appropriate general subject committee (e.g., the CCFAC) on provisions requiring specific technical expertise. Thus when a commodity standard is under development, the commodity committee forwards all proposed food additive provisions to the CCFAC for endorsement.

In March 1991, the FAO/WHO Conference on Food Standards, Chemicals in Foods, and Food Trade was held to review aspects of the work and procedures of Codex, as well as import and export controls that impeded international trade in food (FAO, 1991). The conference recommended that Codex should strengthen the horizontal work of its general subject committees so that matters of general importance, such as labeling, additives, contaminants, and methods of analysis and sampling, would be handled entirely by the appropriate general subject committee. These committees would provide the main source of direction in their areas of expertise and would not be dependent on proposals or provisions put forward by commodity committees. This was seen as necessary if Codex standards were to encompass all foods that were moving in international trade and to provide general guidance and recommendations to promote safe food handling and processing. In response to this conference, CAC shifted its focus from the development of commodity standards to the development of general subject or so-called horizontal standards. As part of this shift in focus, the CCFAC was charged with developing a general standard for the use of food additives, and the various Codex commodity committees were charged with revising existing commodity standards by incorporating changes that reflect the emerging Codex general subject standards.

CONSTRUCTION OF THE GSFA

The raw materials for constructing the GSFA are a food category system, additives for which the Joint FAO/WHO Expert Committee on Food Additives (JECFA) has

---

[1]In the Codex eight-step standard elaboration process, the phrase *proposed draft standard* refers to text under consideration at step 3, the phrase *draft standard* refers to text at step 5, and the phrase *standard* refers to text that has been fully adopted by the CAC and is a Codex standard. This terminology is used throughout this chapter to denote the status of the subject text.

performed a safety review, and food additive use provisions from Codex member states and Codex commodity standards. As of June 1999, JECFA had completed the review of over 900 additives and chemically defined flavors, and Codex had identified 29 additive functional classes (CAC, 1995a).

The food category system is based on one developed by the Confédération des Industries Agro-Alimentaries de la CEE (CIAA) and is intended as a tool to simplify the reporting of food additive uses for constructing the GSFA. It encompasses all foods, including those in which the use of additives is not expected. The food descriptors in the food category system are for foods as marketed. However, they are not intended to be legal product designations, nor are they intended for labeling purposes. The system is hierarchical, meaning that when the use of an additive is permitted in a general food category or subcategory, its use is also allowed in any further subcategories and in individual foods covered by that subcategory.

When reporting food additive uses, member countries are requested to provide the name of the additive, additive functional class, International Numbering System (INS) number, and the maximum level (concentration in mg/kg or mg/l) of use of the additive in the food or food category. This information, along with food additive provisions from Codex commodity standards, is used to construct source worksheets. The worksheets are sorted into two formats. The so-called Schedule 1 or Table 1 format is an alphabetical listing of additives, and the Schedule 2 or Table 2 format is based on food categories. The data in both formats are identical. They differ only in presentation to facilitate ease of use. At the conclusion of the 30th CCFAC (in 1998), the committee had collected additive use information on all additives assigned an acceptable daily intake (ADI) by JECFA.

The data in the source worksheets are compressed by removing all redundant additive usage information as well as the source of the food additive provision (Codex or member state), and by reporting only the highest reported maximum use level for each food or food category. These compressed worksheets are referred to as tables[2] to distinguish them from the raw data compiled in the source worksheets. The 30th CCFAC agreed on the components of the Draft GSFA and the format of Tables 1 and 2.

The CCFAC is committed to ensuring that the additive provisions in the GSFA are compatible with JECFA's ADIs. To accomplish this goal, the CCFAC has agreed to apply the principles of the Budget Method developed in Denmark (Hansen, 1979) as an initial screen to prioritize additives in the draft tables for further consideration. Additives with the highest priority based on the Budget Method will be forwarded to JECFA for evaluation of relevant information about consumer intake of the additive. The 29th CCFAC (in 1997) identified five additives (sulfites, butylated hydroxyanisole [BHA], butylated hydroxytoluene [BHT], *tert*-butylhydroxyquinone [TBHQ], and benzoates), and the 30th CCFAC (in 1998) identified four colors

---

[2]The term *worksheet* is used in this chapter to indicate a compilation of additive use information that has been provided by Codex member states or derived from Codex commodity standards. The term *schedule* refers to data on additive use that is derived from the worksheets, but has been compressed or condensed to reduce redundancies or nonessential information (e.g., country of origin). The term *table* refers to reformatted information in the schedules. *Table* is the term used to refer to the presentation in the current Draft GSFA.

(annatto extract, canthaxanthin, erythrosine, and iron oxides) and referred them to JECFA. The 51st JECFA (in 1998) reviewed national intake data on the five food additives, and the 53rd JECFA (in 1999) reviewed national intake data on the four colors.

The CCFAC has also agreed to apply principles for justifying the technological need for additive use when developing the GSFA. The CCFAC has agreed on the following principles:

- A report of the use of an additive by a national authority is prima facie justification of technological need.
- Technological need may differ from one country to another.
- Technological need should be addressed, whenever possible, through consideration of additive functional classes and not on a case-by-case basis.
- Application of the principles for justifying technological need must be distinct from additive intake assessment.

The CCFAC (CAC, 1998) has also agreed on an approach for resolving questions raised about the technological need and justification for the use of an additive.

CHRONOLOGY OF DEVELOPING THE GSFA

18th CCFAC—1985

The question of whether the CCFAC should provide opinions on the use of food additives other than those included in Codex commodity standards was initially raised by the Codex Secretariat during this session (CAC, 1985). The absence of provisions for the use of additives in nonstandardized foods was also highlighted in a report to Codex by Kermode (1986) in which the author expressed concern about the status and future direction of Codex and its subsidiary bodies.

19th CCFAC—1987

A joint paper (CAC, 1986) prepared by the Codex Secretariat and the Netherlands presented to the CCFAC suggested procedures that could be followed to provide for additive use in foods for which no Codex standards exist. The CCFAC supported the views of the U.S. delegation, which foresaw many difficulties in initiating an exercise to establish provisions for the safe use of additives in nonstandardized foods. Most notable was the need for a consensus on the list of permitted uses of food additives. While member states may have their individual lists, the combined list would consist of numerous and different food types in which each additive could be used. Obtaining a consensus on the technological need for all such uses would be difficult because of differences in regional and cultural dietary habits, and regional public health measures. Moreover, procedures to evaluate the cumulative intake of an additive based on a long list of additive uses were unavailable and would have to be developed (CAC, 1987a).

20th CCFAC—1988

The Codex Secretariat presented a paper (CAC, 1987b) that outlined two approaches for maintaining the up-to-date status of food additive provisions in Codex standards: (1) continue to review additive provisions on a case-by-case basis, or (2) develop general provisions for the use of additives. The CCFAC recognized that the second approach could be extended, at least in principle, to nonstandardized foods. Many delegations expressed concern that the second approach would lead to inadequate control of food additive use and that the standard might permit the use of a number of additives without consideration of their technological need. The CCFAC agreed "that a consultant should prepare a paper on the future activities of the CCFAC in regard to the establishment and regular review of provisions relating to food additives in Codex Standards, and the possible mechanism for establishment of general provisions for the use of food additives in non-standardized foods as a horizontal approach in light of changing requirements in international trade" (CAC, 1988).

21st CCFAC—1989

In response to the 20th CCFAC's recommendation for a paper, the "Denner Report" (Denner, 1989) was delivered to the CCFAC for discussion. The essence of the report is contained in 10 recommendations to CCFAC, FAO, WHO, JECFA, and Codex member states. The following recommendations were directed at the CCFAC.

Recommendation 5

CCFAC should consider in the light of "principles for the safety assessment of food additives and contaminants in food" whether it has sufficient information from JECFA on how to translate ADIs into levels of use in food and drink. If so, then it should produce clear guidelines so that all member nations will understand what factors are, and are not, included in the overall safety factor to ensure, for example, that everyone takes account of special groups in the population (especially children) in the same way. If CCFAC requires further assistance from JECFA then a comprehensive list of clear direct questions should be prepared and forwarded to JECFA (Denner, 1989).

Recommendation 8

CCFAC can never properly carry out its function of endorsing food additive usage in individual foods unless it considers additive usage in all foods. CCFAC should formally take on this task to enable it to serve Codex more effectively. In consequence, the Codex Alimentarius Chapter on Food Additives requires a major revision including a complete restructuring to accommodate provisions for non-standardized foods (Denner, 1989).

Recommendation 10

The CCFAC should adopt the following work plan:

(i) Agree to prepare a new Codex Standard for Food Additives along the lines proposed (paras. 39–40 and Appendix III of CX/FAC 89/16)

(ii) Set up three working parties to deal with different classes of additives to begin work, as a first priority, with additives with an ADI of 10 or less (paras. 41–46 of CX/FAC 89/16)

(iii) Collect usage and intake data for additives with an ADI of 10 or less and, where necessary, prepare a list of restrictions on use taking proper account of technological need and national "styles" of product (paras. 47–51 and 55 of CX/FAC 89/16)

(iv) Redraft the food additive provisions of existing commodity standards to cross refer to the new standard for food additives (paras. 52–54 of CX/FAC 89/16)

(v) As an interim measure, allow the "low priority" additives to be used subject only to GMP, except in those foods from which they would be totally excluded (para. 56 of CX/FAC 89/16) (Denner, 1989).

22nd CCFAC—1990

The CCFAC elected to address Recommendations 5, 8, and 10 of the Denner Report. The CCFAC endorsed Recommendation 5, which was to forward clear, direct questions to JECFA if the CCFAC was unclear about relating toxicologically-derived ADIs to additive use levels in foods.

With respect to Recommendations 8 and 10, the CCFAC agreed to limit, for the time being, its efforts to develop a general standard for food additives to only those additives that had been evaluated by JECFA (CAC, 1990a). The CCFAC also agreed that, in developing a general standard, priority should not be given to any particular group of additives, or to additives with low ADIs, since it was more important to consider the ADI within the context of potential intake. Several delegations expressed their concerns about the complexity of developing a general standard for food additives, especially the need for assessing additive intake on a worldwide basis, and the national nature of many of the nonstandardized foods. Attention also focused on the need for categories of nonstandardized foods. As a starting point, the CCFAC requested the Secretariat to prepare a report on antioxidants and preservatives by grouping together the present Codex uses for these additives in the format proposed in the Denner Report.

23rd CCFAC—1991

The CCFAC considered government comments (CAC, 1991) to the Codex Secretariat's report on antioxidants and preservatives (CAC, 1990b), which was based on food additive provisions derived from Codex commodity standards. To facilitate the development of a general standard for the use of additives, the CCFAC established an ad hoc working group (GSFA WG) with the United States serving as chair and the following terms of reference:

- Establish general principles for the GSFA, including a discussion of its proposed format and scope.
- Elaborate a GSFA that includes all foods and that initially should be restricted to antioxidants and preservatives.
- Complete the document in time for Codex member states to comment prior to the 24th CCFAC.

The GSFA WG met during the 23rd CCFAC. The 19th Session of the CAC (in 1991) agreed: (1) to support the CCFAC's establishment of the GSFA WG under the chair of the United States; and (2) that the CCFAC should prepare a Proposed Draft GSFA based on the Denner Report and the terms of reference for the GSFA WG for circulation and comment at step 3.

24th CCFAC—1992

The 24th CCFAC (CAC, 1993a) agreed that the committee should develop guidelines for translating JECFA ADIs into maximum use levels for food additives, that the GSFA should cover food additive usage in all foods (standardized and nonstandardized), and that the ADI should not be used as the basis for prioritizing additives. A consensus on whether the format of the GSFA should be based on additive functional class (e.g., antioxidants, preservatives, and colors) or on food categories was not reached.

The GSFA WG recommended, and the CCFAC agreed to, the following principles for developing the GSFA:

- The format of the GSFA should be based on functional class titles as provided in the Codex INS list for food additives (CAC, 1995a) and also on food categories initially using the CIAA system.
- The GSFA should include both standardized and nonstandardized foods.
- All additives in the INS list should be included, beginning with those evaluated by JECFA; other food additives should be included only after JECFA completes its safety evaluation.
- The Budget Method (Hansen, 1979) should be used as an initial screen to help establish the maximum level of use for an additive. Codex Guidelines for the Estimation of Food Additive Intake (CAC, 1989) should also be used.
- The GSFA should define foods or food categories where the use of food additives is restricted or prohibited.
- ADIs should not be used to prioritize additives for consideration.

The 24th CCFAC also agreed that the GSFA should have a preamble containing the General Principles for the Use of Food Additives (CAC, 1993a) and refer to appropriate sections of the Codex *Procedural Manual* on the use of food additives (CAC, 1997a). Consensus was not reached on the following:

- substances that may be regarded as foods or food additives according to national legislation
- additives with a long history of safe use in food for which JECFA has not considered or completed a safety evaluation
- application of the principles of technological need and good manufacturing practices (GMPs) in the context of the development of the GSFA

The Codex Secretariat agreed that Circular Letters should be sent requesting comments on the Proposed Draft GSFA, the CIAA list of food categories, and the food categories or food items in which antioxidants or preservatives were not used and where their use was restricted.

25th CCFAC—1993

Prior to this session of the CCFAC, the Codex Secretariat circulated *Proposed Draft Codex General Standard for Food Additives* (CAC, 1992) for comment. This document contained provisions for the use of antioxidants and preservatives based on food additive provisions in Codex commodity standards. The proposed draft standard consisted of five principal sections:

1. Preamble
2. Schedule 1—an alphabetical listing of antioxidants and preservatives by functional class, their maximum use levels and conditions of use in standardized foods, and the applicable guidelines for their use in nonstandardized foods
3. Schedule 2—a listing of the same data in Schedule 1 arranged by food category
4. Annex A—the guidelines for the use of additives in nonstandardized foods, which were based on the Budget Method
5. Annex B—cross-reference listing of Codex standard numbers, Codex standard titles, and CIAA food categories.

The CCFAC's deliberations (CAC, 1993b) on the Proposed Draft GSFA resulted in a number of recommendations for revising its Preamble, Schedule 1, Schedule 2, and Annex A. The most significant revisions to the Proposed Draft Preamble included clarification of which additives and foods the GSFA should cover, how maximum use levels should be established, and criteria for justifying the use of an additive.

The CCFAC forwarded the revised Proposed Draft Preamble to the CAC for adoption at step 5. The 20th Session of the CAC subsequently adopted the Preamble at step 5 (CAC, 1993c).

The 25th CCFAC agreed to revise Schedule 1 and Schedule 2 of the Proposed Draft GSFA using information in Appendix III of ALINORM 93/12A (CAC, 1993d) and to make editorial revisions to the format of the schedules. The committee agreed to request that member states provide information on additive usage levels in foods or food categories, and information justifying the technological need for these levels, with the understanding that this information would be used to revise the schedules, which would be circulated at step 3 for comment in time for the 26th CCFAC.

With regard to Annex A, some delegations expressed their view that the Budget Method was not applicable to Asian countries because it was based on Western dietary habits. In response to these views, the CCFAC agreed that for assessing additive intake, food intake data should be provided whenever possible, but that the Budget Method could be used as a first screening method when no other method was available.

Similarly, some delegations expressed the view that the food category system was based on Western dietary habits and did not adequately account for dietary habits of all geographic and cultural regions of the world. The CCFAC agreed that the food category system should be amended to reflect food categories globally, and member states were requested to provide recommendations for amending the food category system.

26th CCFAC—1994

Discussion of the GSFA at the 26th CCFAC focused on: (1) the section in the Draft Preamble on the Carry-Over Principle,[3] (2) technological justification and need for the use of food additives, (3) further revision of Proposed Draft Schedules 1 and 2, and (4) revision of Proposed Draft Annex A (CAC, 1995b).

The CCFAC agreed to delete from the Draft Preamble two paragraphs on the Carry-Over Principle based on the recommendations of the GSFA WG (CAC, 1995c). In so doing, the Carry-Over Principle as described in the Preamble to the Draft GSFA was no longer the same as that adopted at step 5 by the 20th CAC (in 1993). The CCFAC agreed to forward the revised Preamble to the CAC for adoption at step 8 (CAC, 1995b). The 21st Session of the CAC (in 1995) subsequently adopted the revised Preamble to the GSFA at step 8 (CAC, 1995d).

The CCFAC also had before it revised Proposed Draft Schedule 1 and Proposed Draft Schedule 2 (CAC, 1995e) that included additive usage information on antioxidants, preservatives, and thickeners, with no references to distinguish standardized foods from nonstandardized foods. As for antioxidants and preservatives, the usage information for thickeners was derived from Codex commodity standards. The revised Proposed Draft Schedules also contained information on foods in which the use of antioxidants and preservatives is not permitted by certain national authorities.

The United Kingdom, supported by some delegations, proposed that the GSFA should be presented in three schedules:

1. Schedule 1—foods that may not contain additives of one or more classes except where specifically provided for by Schedule 3

---

[3]The Carry-Over Principle addresses the use of additives in ingredients used to formulate food. For example, the use of an anticaking agent in spice mixes that are used in the preparation of canned beef stew (CAC, 1995a).

2. Schedule 2—food additives generally permitted for use in food in accordance with GMP
3. Schedule 3—food additives permitted for certain uses only

The CCFAC rejected this proposal, noting that the current format, which was previously agreed to by the committee and the CAC, followed a horizontal approach based on food additives and that specific listings as proposed by the United Kingdom could be easily extracted after the standard was published in electronic form. Moreover, the CCFAC noted that the current format of the schedules provided for more effective data collection and should therefore be retained for the present. The CCFAC also agreed that the list of individual foods within food categories would need to be further simplified and that additives with JECFA ADIs of "not specified" should be allowed for use under principles of GMP and based on technological justification and need. The CCFAC, recognizing a need for further discussion of the principles of technological justification and need, agreed to accept the offer of the delegations of Iceland and New Zealand to prepare a discussion paper for the 27th CCFAC in 1995.

The 26th CCFAC concluded its discussion on the GSFA by reaffirming its agreement on the horizontal approach based on the use of food additives in all foods, by agreeing to maintain the two-schedule format, and by calling for further revision based on additional use information. The committee agreed to request all Codex member states to provide the following information for all antioxidants and preservatives evaluated by JECFA:

- categories, subcategories, or specific foods in which antioxidants and preservatives are not permitted
- for each permitted antioxidant and preservative
  — technical effect
  — INS number
  — maximum use level and residual level in food
  — food categories (with food category numbers) in which the additive is used
  — technological justification for use

The request for information was distributed (CAC, 1994a) with responses to be directed to the United States, as chair of the GSFA WG, for compilation into new revised worksheets for consideration at the next CCFAC meeting.

The CCFAC also had before it a paper entitled *Risk Assessment Procedures Used by Codex Alimentarius and Its Subsidiary and Advisory Bodies* (CAC, 1993e), prepared for the 20th Session of the CAC. This paper described how the work of expert bodies, like JECFA, and Codex committees fit into the Codex risk analysis framework. The paper advocated that Codex committees adopt common risk analysis principles and recognize the need for quantitative exposure assessments as part of risk assessment. After discussing this paper, the CCFAC accepted the offer of the delegation from the United Kingdom to prepare a discussion paper on procedures for the evaluation of food additive intake data used in risk analysis, for circulation and comment before the 27th CCFAC.

27th CCFAC—1995

The 27th CCFAC considered

1. revised worksheets for antioxidants and preservatives (CAC, 1995e) based on information provided by Codex member states and provisions contained in Codex commodity standards
2. a revised Proposed Draft Annex A
3. a discussion paper on risk assessment and risk management, prepared by the delegation from the United Kingdom (CAC, 1994b)
4. a discussion paper on the principles of the justification of technological need, prepared by Iceland and New Zealand (CAC, 1995f)

During the GSFA WG's discussion of the revised worksheets for antioxidants and preservatives, many delegations expressed concern about the enormous quantity of information contained in the source worksheets. It was apparent that a means to condense or compress all the reported maximum use levels must be found to proceed with the development of the standard. The GSFA WG considered the use level(s) that would be most appropriate for entry in the next revision of the GSFA. Various delegations argued for the minimum reported, the average value, the median, and the maximum. The GSFA WG concluded, and the CCFAC agreed, that the next version of the proposed draft standard should contain the reported range of use levels for an additive in a given food category. The U.S. delegation volunteered to revise the worksheets by removing redundant information and to implement the other recommendations of the CCFAC in time for its 28th session.

The GSFA WG also discussed the paper on risk assessment and risk management, prepared by the United Kingdom delegation, and offered several recommendations to the CCFAC. The CCFAC reaffirmed its commitment to carry out risk assessment and risk management by developing guidelines, and recommended that JECFA should continue to perform hazard evaluations. The CCFAC agreed to continue its development of risk management procedures with the understanding that further action would be taken when *Application of Risk Analysis to Food Standards Issue, Report of the Joint FAO/WHO Expert Consultation* (WHO/FAO, 1995) became available and its recommendations could be considered. Several delegations expressed their concern that regional and cultural differences in dietary behavior should be taken into account in any method developed to assess additive intake. Most delegations supported the initial use of general screening methods (Budget Method or per capita intake estimates). Others advocated more accurate assessments of additive intake that can account for diverse cultural dietary patterns or dietary habits of special groups (e.g., children and women of childbearing age). The 27th CCFAC did not reach a consensus on how to assess additive intake for purposes of establishing maximum additive use levels in the GSFA. The CCFAC requested that the United Kingdom delegation provide more precise proposals on exposure assessment by using the Budget Method as an initial screening method. The committee also considered a proposal to form a working group on intake assessment under the CCFAC, but no consensus was reached on this proposal. As a result of the CCFAC's discussion of the paper prepared by the United Kingdom, the committee agreed that the Proposed Draft Annex A should also be revised and accepted the offer of the United Kingdom delegation to do so.

The CCFAC's discussion of the paper on technological justification and need for the use of additives centered on the differences in technological needs among countries within the context of Codex's overall objectives of ensuring safety and facilitating international trade. The committee discussed the information that would be sufficient to justify technological need. Some delegations argued that member states should provide the technical data to CCFAC for evaluation of each additive use. Other delegations considered that their responses to the request for additive use information provided prima facie justification for the technological need for that additive class in a particular food category or food item. Several delegations also advocated that no further justification should be required for additive uses contained in existing Codex standards. Other delegations argued that the committee should focus on the technological need through consideration of additive classes rather than on a case-by-case approach. The delegations from Iceland and New Zealand accepted the CCFAC's request that they revise the discussion paper by elaborating on these aspects, and forward more precise proposals for how technological need should be considered in the development of the GSFA.

Finally, the CCFAC agreed to request that member states provide use information on three additional functional classes of additives: (1) thickeners, (2) stabilizers, and (3) sweeteners. Once again, the U.S. delegation offered to compile the information and construct source worksheets for these additive functional classes.

28th CCFAC—1996

The 28th CCFAC was held in Manila, the Philippines, at the invitation of the governments of the Netherlands and the Philippines. The Netherlands chaired the meeting. The GSFA WG addressed several issues:

1. compression of the additive use information on antioxidants and preservatives (CAC, 1995g)
2. revision of the food category system (CAC, 1995h)
3. technological justification and need
4. revision of the Preamble
5. revision of the Proposed Draft Annex A
6. worksheets on stabilizers, thickeners, and sweeteners

The GSFA WG considered the compressed source worksheets (schedules) for antioxidants and preservatives prepared by the U.S. delegation. That delegation provided an overview of the process for compressing the worksheet data into schedules and emphasized that the schedules are intended as tools for formatting the final standard and for establishing maximum additive use levels. The schedules were constructed by using a revised food categorization system that attempted to account for previous concerns regarding the applicability of the food category system to non-Western diets. During the plenary session, several delegations again asserted that the revised food category system still did not adequately reflect non-Western diets. The committee reaffirmed its commitment that the categorization system should include all foods in international trade and requested that specific recommendations be submitted to amend the food category system accordingly. Subsequently, recommendations were submitted and the food category system was further revised (CAC, 1996a).

No discussion paper on the application of the principles of technological justification and need was available for discussion. During the GSFA WG meeting, the delegate from New Zealand presented a proposed stepwise procedure for applying principles for justifying technological need toward the development of the GSFA. There was general agreement by the GSFA WG that the declared use of an additive in a particular food category or food item by a member state is prima facie justification for technological need. The 28th CCFAC agreed that New Zealand, Australia, and Iceland would further elaborate the application of the principles of technological justification and need by drafting a discussion paper expanding on the stepwise procedure discussed during the GSFA WG.

The need for further revision of the Preamble was also considered by the GSFA WG. The 21st CAC requested that provisions in the Preamble referring to the Carry-Over Principle be revisited. The committee agreed to forward to the CAC amendments to the Preamble to include new language on the Carry-Over Principle.

The GSFA WG also recognized that the Preamble contained no instructions or procedures for amending the food additive provisions of the GSFA once the standard was fully elaborated. The CCFAC accepted the offer of the U.S. delegation to propose revisions to the Preamble to provide for the amendment of the GSFA.

The CCFAC also accepted the offer of the Belgian delegation, in cooperation with the CIAA, to propose amendments to the Preamble to describe more thoroughly the principles and application of the food category system.

In response to a request by the 27th CCFAC, the United Kingdom delegation presented a paper proposing a four-tiered approach for assessing additive intake (CAC, 1996b). The proposed tiers ranged from simple to complex and resource-intensive. The four tiers were:

1. Tier 1—Budget Method
2. Tier 2—a "reverse" budget method
3. Tier 3—model diet that incorporates actual food consumption data and that, in principle, is amenable to modification to account for regional and cultural diets (Unit Quantity Diet)
4. Tier 4—national food intake data

The CCFAC, in principle, supported this tiered approach, but expressed reservations about the limitations of the methods. The committee accepted the offer of the United Kingdom delegation to further elaborate the paper on intake assessment methods and to revise the Proposed Draft Annex A of the GSFA in light of the approach. The CCFAC also accepted the offer of the United Kingdom delegation to provide intake assessment data on antioxidants, preservatives, and if possible, stabilizers, thickeners, and sweeteners in a timely manner to allow for further elaboration of the standard prior to the 29th CCFAC.

The CCFAC agreed to request use information on the following additive functional classes: colors, color retention agents, bulking agents, and emulsifiers. The committee again accepted the U.S. delegation's offer to construct source worksheets for these additive classes in time for the 29th CCFAC. In addition, with the United Kingdom delegation's commitment to deliver intake assessments in a timely manner, the U.S. delegation offered to revise the Proposed Draft Schedules for antioxidants and preservatives, and if possible, stabilizers, thickeners, and sweeteners in time for distribution prior to the 29th CCFAC. The committee accepted the U.S. delegation's offer and requested that, in addition to the revised Proposed Draft Schedules, the U.S. delegation provide an explanation of how the proposed draft schedules were constructed from the source worksheets.

The observer from the European Community, as well as some member states, suggested that the committee should consider all additives with JECFA ADI "not specified" in a separate list in accordance with the principles of GMP as well as a list of food categories or foods where additives were not allowed. The CCFAC did not support this proposal.

29th CCFAC—1997

The CCFAC was at a critical juncture in the development of the GSFA. For the first time, the CCFAC had produced a work product containing provisions for the general use of additives in food that could be forwarded to the CAC for adoption.

The CCFAC considered the following documents related to the GSFA:

1. revised Preamble (CAC, 1996c)
2. Proposed Draft Schedules for antioxidants, preservatives, stabilizers, thickeners, and sweeteners (CAC, 1996d)
3. revised Proposed Draft Annex A (CAC, 1996e)
4. consideration of technological justification and need (CAC, 1997b)
5. source worksheets for colors, color retention agents, bulking agents, and emulsifiers (CAC, 1996f).

The revised Preamble contained new sections that addressed the Carry-Over Principle to make it consistent with Section 5.3, Volume 1A of the Codex Alimentarius (CAC, 1995a), the principles of the food category system used in the GSFA, procedures and requirements for amending the GSFA once the standard is fully elaborated, and editorial changes.

The proposed draft schedules for antioxidants/preservatives, stabilizers/thickeners, and sweeteners were presented in three parts: Proposed Draft Schedule 1 (organized by alphabetical listing of food additives), Proposed Draft Schedule 2 (organized by food category), and Proposed Draft Schedule 3 (a listing of food categories and food items in which specific additives are not allowed by national authorities). In addition to these schedules, the U.S. delegation presented a paper describing how Schedule 1 and Schedule 2 were constructed from the information submitted by member states, the food additive provisions in Codex commodity standards, and the intake assessment provided by the United Kingdom delegation.

In discussing these schedules, the observer from the European Community again proposed that the committee place all additives with JECFA ADIs of "not specified" or "not limited" (i.e., nonnumerical) in a separate schedule and endorse their use in foods in general in accordance with GMP, along with a list of food categories or foods where these additives were not allowed or were further restricted. The CCFAC endorsed this approach and established a list (CAC, 1997c) of approximately 170 additives for use in food in general in accordance with GMP along with an annex containing a list of food categories or individual foods in which the use of these additives was not allowed or was restricted. This annex of food categories (CAC, 1997d) is based on a list that is currently in effect in the European Union (EU). As corollaries to this decision, the committee agreed to cease further development of Schedule 3 formatted information and to delete all food additive provisions in Proposed Draft Schedule 1 and Proposed Draft Schedule 2 reported in food category 0.0 (Foods in General). The GMP list and its annex were forwarded by the CCFAC to the 22nd CAC at step 5 with a recommendation for adoption at step 8. The 22nd CAC subsequently adopted the GMP list of additives at step 8, and adopted the Annex on an interim basis at step 8 with a request that the CCFAC review the Annex and report its findings to the next session of the CAC (CAC, 1997f, paras. 57 and 59). The CCFAC agreed to forward provisions for the use of additives with numerical JECFA ADIs to the 22nd CAC at step 5. The 22nd CAC adopted these provisions at step 5 (CAC, 1997f, para. 114).

The revised Proposed Draft Annex A described a three-tiered approach for assessing additive intake for purposes of establishing maximum use levels in the GSFA (CAC, 1996e). Tier 1 was based on a modified Budget Method, Tier 2 described a model diet that was intended to be tailored to different cultural and regional dietary habits (Unit Quantity Diet), and Tier 3 discussed the need for national intake data. An accompanying paper described the application of this tiered approach to the compressed source worksheets for antioxidants, preservatives, stabilizers, thickeners, and sweeteners; and proposed modifications to the application of the tiered approach for the development of the GSFA. Results from the application of the methods described in the revised Annex A to the data in the compressed source worksheets were used to construct Proposed Schedule 1 and Proposed Draft Schedule 2 for these additive functional classes. The approach described in this paper was discussed by the GSFA working group and the CCFAC plenary. The committee agreed that the proposed tiered approach, while helpful, was not adequate for furthering the development of the GSFA. Therefore, the CCFAC agreed that the Proposed Draft Annex A should be revised to incorporate only the modified Budget Method and to refer additives to JECFA for the evaluation of data on probable human exposure, as appropriate.

The paper proposing a stepwise procedure for the evaluation of technological need was based on the following decisions of the 26th and 27th CCFAC:

1. recognition that technological need may differ from one country to another
2. recognition that technological need should be addressed, whenever possible, through consideration of additive functional classes and not on a case-by-case basis
3. maintenance of a clear distinction between the justification of technological need and additive intake assessment

The paper recommended that the first premise for justifying the need for the use of an additive is whether a national authority has reported such a use for an additive. If the initial intake assessment indicates that intake of the additive may exceed the ADI, then a closer examination of the technological need of the proposed maximum use levels for the additive in all food categories should be undertaken with a view to refining the intake assessment, and subsequent refinement of the maximum use levels. The paper envisioned that this process would continue unless the exposure assessment could be shown to be compatible with the ADI. The committee agreed to request comments on this paper for consideration by the 30th CCFAC.

The worksheets for colors, color retention agents, bulking agents, and emulsifiers were prepared in response to a request to member states for additive use information and additive provisions derived from Codex commodity standards.

The 29th CCFAC agreed to the following:

- Forward the amendments to the Preamble to the CAC for adoption at step 8 (CAC, 1997e).
- Forward to the CAC at step 5, with a recommendation for adoption at step 8, a list of 170 additives with JECFA ADIs "not specified" or "not limited" that are permitted for use in food in general, unless otherwise specified, in accordance with GMP, together with a list of food categories and food items in which the use of these additives is prohibited or further restricted (CAC, 1997c and d).
- Remove the Foods in General category from the proposed draft schedule for antioxidants, preservatives, stabilizers, thickeners, and sweeteners with numerical JECFA ADIs and forward the revised proposed draft schedules to the 22nd CAC for adoption at step 5 (CAC, 1997e).
- Identify food categories in which the use of food additives is inappropriate; current Schedule 3 is not suitable for this purpose.
- Discontinue the tiered approach described in the Proposed Draft Annex A and request the Danish delegation, assisted by the French and United Kingdom delegations, to revise the Proposed Draft Annex A to contain a new version of the Budget Method for circulation prior to the 30th CCFAC.
- Apply the Budget Method to Proposed Draft Schedule 1 and Proposed Draft Schedule 2 of the GSFA to prioritize additives with numerical ADIs and refer an appropriate number of additives for JECFA evaluation of national data on probable human exposure.
- Request additional comments on the discussion paper on technological justification and need prepared by Australia, Iceland, and New Zealand.
- Request the U.S. delegation to compress the worksheets for colors, color retention agents, bulking agents, and emulsifiers prior to the 30th CCFAC.
- Request additive use information on all remaining classes of additives, except flavorings, with numerical JECFA ADIs.

49th JECFA—June 1997

In response to the request from the 29th CCFAC to evaluate additive intake information, a framework paper describing the types of data JECFA would need to evaluate probable exposure to food additives was prepared and discussed by the 49th JECFA. As a result of these deliberations, a data call was issued requesting national intake data for sulfites, TBHQ, BHT, BHA, and benzoates be submitted to the JECFA Secretariat for evaluation at the 51st JECFA in June 1998.

30th CCFAC—1998

The CCFAC had now collected use information for all additives assigned an ADI by JECFA. These data were collected based on all additive functional classes (except flavoring agents).

The following documents were discussed by the 30th CCFAC:

- draft schedules for antioxidants, preservatives, stabilizers, thickeners, and sweeteners (step 5)
- proposed draft schedules for colors, color retention agents, bulking agents, and emulsifiers (step 3)
- worksheet data for acidity regulators, anticaking agents, antifoaming agents, firming agents, flavor enhancers, flour treatment agents, foaming agents, glazing agents, humectants, propellants, and raising agents (step 3)
- proposed draft revised Annex A (*Guidelines for the Estimation of Appropriate Levels of Use of Food Additives*) (step 3) (CAC, 1997g)

The GSFA WG considered the following issues: (1) the final format of the GSFA; (2) resolution of questions of the technological need and justification for specific uses of additives; and (3) the use of Annex A to screen proposed maximum use levels in the Draft GSFA.

In response to previous discussions by the GSFA working group and in recognition of the large quantity of information that the committee had collected and compiled on additive uses, the U.S. delegation, as chair of the GSFA WG, proposed a new format for the GSFA. The GSFA WG proposed additional modifications and the 30th CCFAC subsequently agreed to the following components for the GSFA:

Preamble
   — Annex A to the Preamble (Guideline for the Development of Maximum Levels of Use for Food Additives with Numerical ADIs)
   — Annex B to the Preamble (GSFA Food Category System)
   — Annex C to the Preamble (Cross-Reference to Codex commodity standards and the GSFA Food Category System)
Index
   — List A: JECFA "Approved" Food Additives with ADIs and INS Numbers (including synonyms and sorted alphabetically by additive)
   — List B: JECFA "Approved" Food Additives with ADIs and INS Numbers (including JECFA review date and meeting number and sorted by INS number)
- Table 1: Additives Permitted for Use Under Specified Conditions in Certain Food Categories or Individual Food Items

- Table 2: Food Categories or Individual Food Items in Which the Use of Food Additives Is Permitted under Specified Conditions
- Table 3: Additives with Nonnumerical ADIs Permitted for Use in Food in General in Accordance with GMP Unless Otherwise Specified
  — Annex to Table 3: Food Categories and Individual Food Items Where the Use of Food Additives with Good Manufacturing Practice Limitations on Use Is Not Allowed or Restricted

In the new format, Table 1 (sorted alphabetically by additive) corresponds to the previous Schedule 1 format, and Table 2 (sorted by food category) corresponds to the previous Schedule 2 format. Similarly, the additive use information in Table 1 and Table 2 is identical; the tables differ only in their presentation. The 30th CCFAC also agreed that an additive's technical effects would be listed in a separate subhead in Table 1 and would not be listed in Table 2. In addition, for additives that share the same JECFA ADI (i.e., grouped additives, such as phosphates and adipates), there would be one entry only in a specific food category. Thus, for a given additive or grouped additive, these changes eliminated multiple entries in individual food categories. There is only one maximum use level for an additive in a specific food category. Finally, the 30th CCFAC agreed that the category entitled Foods in General would be removed from the draft tables. The U.S. delegation, as chair of the GSFA WG, accepted the CCFAC's request to revise the Draft GSFA for circulation and comment in time for the next CCFAC.

The 30th CCFAC also agreed to forward the step 3 provisions in the GSFA to the Codex Executive Committee for adoption at step 5. The 45th Session of the Codex Executive Committee (in 1998) subsequently adopted these provisions at step 5. Thus, all of the food additive provisions in the Draft Table 1 and Draft Table 2 were at step 5.

In response to a proposal from the Australian delegation, which was recommended by the GSFA WG, the 30th CCFAC agreed to the following procedure for resolving questions of technological need and justification for food additive provisions in the Draft GSFA:

- Establish that at least two Codex member states permit the use of the additive up to the maximum level proposed in Table 1 and Table 2 in foods representative of the category. This establishes that trade may occur in the food containing the additive.
- Establish whether the maximum level proposed is limited to an obscure or unrepresentative food. If so, consideration may be given to recognizing that food and the level of additive use as a specific entry in the GSFA, and identifying a more representative level for the category as a whole.
- Use square brackets as appropriate, where Codex member states continue to express concern about the proposed maximum levels.
- Circulate the revised Draft Table 1 and Draft Table 2 for comments.
  — If a member state considers the proposed level of use too high, data should be presented to demonstrate that the use level presents a risk to public health, may lead to consumer deception about the nature of the food, or is otherwise technologically unnecessary.

— If a member state wishes to support a draft maximum use level that has been identified as being of concern by other Codex members states, data should be presented to demonstrate that the product could not be made to a satisfactory quality using a lower level of additive or alternative additives that are permitted in the GSFA.

The CCFAC also agreed to return the revised Proposed Draft Annex A to step 2 for revision by the delegation of Denmark and to request JECFA review national intake data on annatto extract, canthaxanthin, erythrosine, and iron oxides.

In discussing the Annex to Table 3, the JECFA Secretariat noted that several additives listed in Table 3 had not been evaluated by JECFA. The 30th CCFAC accordingly agreed to recommend the deletion of these substances from the GSFA. In addition, the JECFA Secretariat clarified the ADI status of some other additives in Table 3, and the 30th CCFAC agreed to amend the table accordingly.

51st JECFA—1998

In response to a request from the 29th CCFAC, JECFA reviewed national intake data on benzoates, BHA, BHT, sulfites, and TBHQ. The 51st JECFA concluded that mean intake estimates by member states—based on GSFA maximum reported use levels and the range of foods in which the additives are allowed integrated with national food consumption data—exceed the ADI for all five food additives. The mean intake estimates based on nationally approved uses and food consumption data, however, did not exceed the ADI. JECFA identified several provisions for the use of these additives in the Draft GSFA that may significantly contribute to intake and recommended that CCFAC review these provisions.

31st CCFAC—1999

For the first time, the CCFAC had before it the newly formatted Draft GSFA. The revised Draft Table 1 and Draft Table 2 at step 5 contained a compilation of additive use information for all additive functional classes of those additives assigned a JECFA ADI (CAC, 1998). In addition, a revised Proposed Draft Annex A to the GSFA Preamble at step 3 was available for discussion, along with a report by the Danish delegation on the application of the revised Proposed Draft Annex A to the food additive provisions in the Draft Table 1 and Draft Table 2.

At the recommendation of the GSFA WG, the CCFAC endorsed the Draft Proposed Annex A with revisions and forwarded it to the CAC with a recommendation for adoption at step 5.

The GSFA WG focused its discussion on the additives identified by the Annex A screen as presenting no possible concern with respect to exceeding the JECFA ADI. At the recommendation of the GSFA WG, the CCFAC agreed to endorse provisions for the use of a number of these additives and forward them to the CAC with a recommendation for final adoption at step 8 (CAC, 1999). Some Codex member states expressed reservations with respect to forwarding only GMP limitations for additives assigned numerical ADIs by JECFA. Other Codex member states noted that the GMP limitations reflect national practices because, in the construction of the Draft Table 1 and Draft Table 2, preference was always given to numerical use levels over just GMP limitations. Therefore, it would be inappropriate for Codex to require establishing numerical limits for additives in cases where national authorities had not found them necessary to ensure the safe use of the additive.

The GSFA WG was unable to complete its discussion of the five additives referred to by the 51st JECFA for review of national intake data. The 32nd CCFAC (in 2000) is expected to continue its discussion of these additives.

As a result of the GSFA WG's discussion of the provisions in the Draft Table 1 and Draft Table 2 and at the suggestion of the observer from the European Commission, the 31st CCFAC agreed that a small intersession working group, charged with reviewing the draft tables, would be useful. The purpose of this working group would be to perform a quality control check to identify errors and to confirm the technological need of the remaining additives. Delegations from Australia, Brazil, the EU, Japan, the Republic of South Africa, and the United States agreed to participate as regional representatives of their respective continents.

The 31st CCFAC also discussed the Annex to Table 3 and agreed to forward to the CAC for adoption at step 8 a revised title to the Annex (Food Categories or Individual Food Items Excluded from the General Conditions of Table 3) and to incorporate additional explanatory text clarifying the intent of the Annex. The CCFAC also recommended adding food categories corresponding to wine and fruit juices to the Annex to Table 3.

The 31st CCFAC also agreed to circulate for comment proposed draft revisions to the Preamble at step 3 to clarify that Table 1, Table 2, and Table 3 do not refer to the use of substances as processing aids. In addition, the committee agreed to circulate proposed draft revisions to the Preamble emphasizing that the lack of a reference to a particular additive in the GSFA does not imply that the additive is unsafe.

CURRENT STATUS OF THE GSFA

At the completion of the 31st CCFAC and the 23rd CAC (in 1999), the status of the GSFA was as follows:

- The Preamble had been adopted at step 8.
- The Draft Annex A to the Preamble containing a description of the application of the Budget Method to the GSFA had been adopted at step 5.
- Table 1 and Table 2 now contain the provisions for the use of 26 additives in specific food categories that had been adopted at step 8.
- Draft Table 1 and Draft Table 2 now contain, at step 6, the provisions for the use of all remaining additives with numerical JECFA ADIs.
- Table 3, which contains a list of approximately 170 food additives for use in foods in general in accordance with GMP, along with a list of food categories or food items excluded from the general conditions of Table 3, had been adopted at step 8.

Draft Table 1 and Draft Table 2 of the GSFA will be revised and circulated for comment in time for the 32nd CCFAC. The revised Draft Table 1 and Draft Table 2 will be amended in the following manner:

1. Comments provided by Codex member states in response to CX/FAC 99/6 (CAC, 1998) will be included as appropriate.
2. Provisions for the use of additives listed in Table 3 in food categories, listed in the Annex to Table 3, will be added. These provisions will reflect uses provided by Codex member states and provisions in Codex commodity standards.

FUTURE WORK

The CCFAC has made significant progress on the elaboration of the GSFA. The 32nd CCFAC (in 2000) will consider the following:

1. the proposed revisions to the Preamble to the GSFA
2. the revised Draft Table 1 and Draft Table 2
3. the quality control recommendations of the intersession working group composed of regional representatives, which should allow the 32nd CCFAC to advance additional additive provisions to step 8
4. written comments from Codex member states on the recommendation of the 51st JECFA that CCFAC review specific provisions for the use of benzoates, BHA, BHT, sulfites, and TBHQ; and the recommendations of the 53rd JECFA based on its evaluation of national intake assessment data on annatto extract, canthaxanthin, erythrosine, and iron oxides
5. proposed revisions to the food category system (Annex B to the Preamble)

REFERENCES

Codex Alimentarius Commission. (1985). *Should CCFA Express Opinions on Food Additives Other Than Those Included in Codex Commodity Standards?* CX/FA 85/16. Rome: Codex Alimentarius Commission.

Codex Alimentarius Commission. (1986). *Procedures To Be Followed by the CCFA To Provide for the Use of Food Additives in Foods for Which There Are No Codex Standards.* CX/FA 87/19-Add.1. Rome: Codex Alimentarius Commission.

Codex Alimentarius Commission. (1987a). *Report of the 19th Session of the Codex Committee on Food Additives.* ALINORM 89/12A, paras. 286–269. Rome: Codex Alimentarius Commission.

Codex Alimentarius Commission. (1987b). *Regular Reviews of Food Additive Provisions in Codex Standards.* CX/FA 88/10-Part 1. Rome: Codex Alimentarius Commission.

Codex Alimentarius Commission. (1988). *Report of the 20th Session of the Codex Committee on Food Additives.* ALINORM 89/12. Rome: Codex Alimentarius Commission.

Codex Alimentarius Commission. (1989). Guidelines for the Simple Evaluation of Food Additive Intake. In *Codex Alimentarius*, 2nd ed. Vol. XIV, Ed. 1, Supplement 2. Rome: Codex Alimentarius Commission.

Codex Alimentarius Commission. (1990a). *Report of the 22nd Session of the Codex Committee on Food Additives and Contaminants.* ALINORM 91/12, paras. 31–37. Rome: Codex Alimentarius Commission.

Codex Alimentarius Commission. (1990b). *Proposals for General Provisions for the Use of Food Additives in Standardized and Non-standardized Foods.* CL 1990/26-FAC. Rome: Codex Alimentarius Commission.

Codex Alimentarius Commission. (1991). *Report of the 19th Session of the Codex Alimentarius Commission.* ALINORM 91/40. Rome: Codex Alimentarius Commission.

Codex Alimentarius Commission. (1992). *Proposed Draft Codex General Standard for Food Additives (GSFA).* CL 1992/18-FAC. Rome: Codex Alimentarius Commission.

Codex Alimentarius Commission. (1993a). *Report of the 24th Session of the Codex Committee on Food Additives and Contaminants.* ALINORM 93/12. Rome: Codex Alimentarius Commission.

Codex Alimentarius Commission. (1993b). *Report of the 25th Session of the Codex Committee on Food Additives and Contaminants.* ALINORM 93/12A. Rome: Codex Alimentarius Commission.

Codex Alimentarius Commission. (1993c). *Report of the 20th Session of the Codex Alimentarius Commission.* ALINORM 93/40, paras. 212–215. Rome: Codex Alimentarius Commission.

Codex Alimentarius Commission. (1993d). Food categories where preservatives and/or antioxidants are/are not used, Appendix III. In *Report of the 25th Session of the Codex Committee on Food Additives and Contaminants.* ALINORM 93/12A. Rome: Codex Alimentarius Commission.

Codex Alimentarius Commission. (1993e). *Risk Assessment Procedures Used by the Codex Alimentarius Commission and Its Subsidiary and Advisory Bodies.* ALINORM 93/37. Rome: Codex Alimentarius Commission.

Codex Alimentarius Commission. (1994a). *Request for Comments and Information on the Use of Antioxidants and Preservatives in Foods.* CL 1994/11-FAC. Rome: Codex Alimentarius Commission.

Codex Alimentarius Commission. (1994b). *Codex Risk Assessment and Management Procedures: The Translation of Advice from JECFA into Codex General Standards for Food Additives and Contaminants.* CX/FAC 95/3. Rome: Codex Alimentarius Commission.

Codex Alimentarius Commission. (1995a). *Codex Alimentarius*, 2nd ed. Vol. 1A, Section 5. Rome: Codex Alimentarius Commission.

Codex Alimentarius Commission. (1995b). *Report of the 26th Session of the Codex Committee on Food Additives and Contaminants.* ALINORM 95/12. Rome: Codex Alimentarius Commission.

Codex Alimentarius Commission. (1995c). Report of the ad hoc Working Group on the General Standard for Food Additives. Conference Room Document 1. *Report of the 26th Session of the Codex Committee on Food Additives and Contaminants.* ALINORM 95/12. Rome: Codex Alimentarius Commission.

Codex Alimentarius Commission. (1995d). *Report of the 21st Session of the Codex Alimentarius Commission.* ALINORM 95/37. Rome: Codex Alimentarius Commission.

Codex Alimentarius Commission. (1995e). *Consideration of Revised Schedules 1, 2, and Annex A of the Proposed Draft Codex General Standard for Food Additives,* and *Government Comments.* CX/FAC 95/4, CX/FAC 95/4-Add.1. Rome: Codex Alimentarius Commission.

Codex Alimentarius Commission. (1995f). *Consideration of Technological Justification and Need for the Use of Food Additives.* CX/FAC 95/5. Rome: Codex Alimentarius Commission.

Codex Alimentarius Commission. (1995g). *Development of Compressed Worksheet Formats for Antioxidants and Preservatives for the Proposed Draft Codex General Standard for Food Additives.* CX/FAC 96/7. Rome: Codex Alimentarius Commission.

Codex Alimentarius Commission. (1995h). *Revision of the CIAA (Confederation of the Food and Drink Industries of the EEC) Food Categories for Development of a Codex Food Identification System (CFIS) in Regard to the Proposed Draft Codex General Standard for Food Additives.* CX/FAC 96/10. Rome: Codex Alimentarius Commission.

Codex Alimentarius Commission. (1995i). *Codex Alimentarius,* 2nd ed. Vol. 5A. Rome: Codex Alimentarius Commission.

Codex Alimentarius Commission. (1995j). *Codex Alimentarius,* 2nd ed. Vol. 7. Rome: Codex Alimentarius Commission.

Codex Alimentarius Commission. (1996a). *Part I: Request for Comments and Information on the Use of Colours, Colour Retention Agents, Bulking Agents, and Emulsifiers in Foods and Part II: Codex Food Categorisation System (CFCS) for the General Standard for Food Additives (GSFA).* CL 1996/14-FAC. Rome: Codex Alimentarius Commission.

Codex Alimentarius Commission. (1996b). *Codex Risk Assessment and Management Procedures: Proposed Exposure Assessment Methods in Support of the Codex General Standard for Food Additives.* CX/FAC 96/6. Rome: Codex Alimentarius Commission.

Codex Alimentarius Commission. (1996c). *Consideration of the Proposed Draft Revised Preamble to the Codex General Standard for Food Additives.* CX/FAC 97/6. Rome: Codex Alimentarius Commission.

Codex Alimentarius Commission. (1996d). *Consideration of the Proposed Draft Schedules for Antioxidants/Preservatives, Stabilizers/Thickeners, and Sweeteners.* CX/FAC 97/7. Rome: Codex Alimentarius Commission.

Codex Alimentarius Commission. (1996e). *Consideration of the Proposed Draft Revised Annex A.* CX/FAC 97/9. Rome: Codex Alimentarius Commission.

Codex Alimentarius Commission. (1996f). *Consideration of the Worksheets for Colours/Colour Retention Agents, Bulking Agents, and Emulsifiers in Regard to the Codex General Standard for Food Additives.* CX/FAC 97/8. Rome: Codex Alimentarius Commission.

Codex Alimentarius Commission. (1997a). *Codex Alimentarius Commission Procedural Manual,* 10th ed. Rome: Codex Alimentarius Commission.

Codex Alimentarius Commission. (1997b). *Consideration of Technological Justification and Need for the Use of Food Additives.* CX/FAC 97/10. Rome: Codex Alimentarius Commission.

Codex Alimentarius Commission. (1997c). *General Standard for Food Additives: Draft Schedules of Additives Permitted for Use in Food in General, Unless Otherwise Specified, in Accordance with GMP.* ALINORM 97/12A, Appendix IV. Rome: Codex Alimentarius Commission.

Codex Alimentarius Commission. (1997d). *Food Categories or Individual Food Items Where the Use of Food Additives with Good Manufacturing Practice Limitations on Use Are Not Allowed or Restricted.* ALINORM 97/12A, Annex to Appendix IV. Rome: Codex Alimentarius Commission.

Codex Alimentarius Commission. (1997e). *Report of the 29th Session of the Codex Committee on Food Additives and Contaminants.* ALINORM 97/12A. Rome: Codex Alimentarius Commission.

Codex Alimentarius Commission. (1997f). *Report of the 22nd Session of the Codex Alimentarius Commission.* ALINORM 97/37. Rome: Codex Alimentarius Commission.

Codex Alimentarius Commission. (1997g). *Proposed Draft Revised Annex A.* CX/FAC 98/9. Rome: Codex Alimentarius Commission.

Codex Alimentarius Commission. (1998). *Draft Codex General Standard for Food Additives: Revised Tables 1, 2, and 3 (Including Annex to Table 3).* CX/FAC 99/6. Rome: Codex Alimentarius Commission.

Codex Alimentarius Commission. (1999). *Report of the 31st Session of the Codex Committee on Food Additives and Contaminants.* ALINORM 99/12A. Rome: Codex Alimentarius Commission.

Denner, W. H. B. (1989). *Future Activities of the Committee in Regard to the Establishment and Regular Review of Provisions Relating to Food Additives in Codex Standards and Possible Mechanisms for the Establishment of General Provisions for the Use of Food Additives in Non-Standardized Foods.* CX/FAC 89/16. Rome: Codex Alimentarius Commission.

Food and Agriculture Organization. (1991). *Report of the FAO/WHO Conference on Food Standards, Chemicals in Food and Food Trade.* Rome: Food and Agriculture Organization.

Hansen, S. C. (1979). *Conditions for Use of Food Additives Based on a Budget for an Acceptable Daily Intake. J Food Protec* 42, 429–434.

Kermode, G. O. (1986). *The Future Direction of the Work of the Joint FAO/WHO Food Standards Programme.* CX/EXEC 86/33/CRD 1. Rome: Codex Alimentarius Commission.

World Health Organization/Food and Agriculture Organization. (1995). *Application of Risk Analysis to Food Standards Issues, Report of the Joint FAO/WHO Expert Consultation.* Geneva, Switzerland, March 13–17, 1995. WHO/FNU/FOS/95.3. Geneva, Switzerland: World Health Organization.

Chapter 12

Development of the Codex Standard for Contaminants and Toxins in Food

Torsten Berg

INTRODUCTION

"Je vis de bonne soupe et non du beau langage." This quotation from Molière's Les Femme Savantes introduces the paper presented to the 21st Session of the Codex Committee for Food Additives and Contaminants (CCFAC) in 1989 in The Hague by the independent consultant W.H.B. Denner (1989). That paper served as the basis for the committee's decisions that eventually will lead to a Codex General Standard for Food Additives (see Chapter 11). The same words and the ideas of Dr. Denner provided inspiration to the work leading to the Codex General Standard for Contaminants and Toxins in Food (GSCTF).

During the 23rd session of the CCFAC in 1991, the delegation of the United Kingdom suggested, and other delegations supported, that there was a need for developing a Codex philosophy on contaminants. The chair, C.G.M. Klitsie, citing that this would be an important issue at the Food and Agriculture Organization/World Health Organization (FAO/WHO) Conference on Food Standards, Chemicals in Food and Trade to be held in Rome later in 1991, invited delegates to express their views at that conference and then discuss the issue at the next CCFAC (CAC, 1991). Following a series of comprehensive discussions of a wide range of individual contaminants later in this meeting, the committee accepted an offer from the delegations of Denmark and the Netherlands to develop a paper on philosophy and procedure concerning the setting of Codex maximum and guideline levels for contaminants for discussion at the next CCFAC session.

Having discussed the Denner paper both in 1989 and 1990 and subsequently taken action on several of its recommendations, in 1991 the committee decided to

---

Acknowledgments: Dr. David Kloet of the Netherlands together with the author developed the papers that eventually built the GSCTF. The author will forever remain grateful for that partnership. CCFAC Chairs Mrs. C.G.M. Klitsie, Mr. Hans van der Kooi, and Mr. Edwin Hecker gave, in turn, their invaluable support and inspiration to the project. Dr. David Byron of the FAO/WHO Joint Secretariat in Rome kept us in line, as well as supported and encouraged us when we needed it.

develop a draft General Standard for Food Additives including all foods (i.e., a horizontal standard for food additives). It was consequently not surprising that the idea emerged to develop one General Standard on Contaminants, which would lay down provisions for all contaminants important for the health of the consumer as well as for international trade in food.

Only six years later, in 1997, the Codex Alimentarius Commission (CAC), which is the superior body of the Codex Alimentarius, accepted the CCFAC proposal for a GSCTF at the final step 8, in the form of a Preamble with five annexes. The five annexes cover, respectively:

1. Criteria for the Establishment of Maximum Limits (MLs) in Food
2. Procedure for Risk Management Decisions
3. Format of the Standard
4. Annotated List of Contaminants and Toxins
5. Food Categorization System To Be Used in the GSCTF

The GSCTF, however, does not contain figures pertaining to the MLs for contaminants and toxins in the various food groups. The MLs are presently under development by the CCFAC for the contaminants included in the GSCTF.

Before plunging into how the GSCTF was developed, it may be worthwhile to present the framework of the Codex system. The Codex Alimentarius was established in 1962 under the FAO/WHO Food Standards Programme. In the early years, Codex focused on developing vertical standards (i.e., standards for individual food products or commodities). For that purpose, the CAC, which meets biennially, established the commodity committees dealing with these subjects (e.g., Codex Committee for Fish and Fishery Products [CCFFP]). The general subject committees were established to deal with food problems in a more general way, like the CCFAC. Most committees meet annually in a host country that also provides a chair for the meeting and committee Secretariat. Toward 1980 it became increasingly clear that it would be difficult to establish worldwide standards for all foods moving in international trade and that the work in many commodity committees was about to be completed. The focus then shifted toward the general standards committees, and many commodity committees were adjourned *sine die*.

When a draft standard is first accepted by a committee, often at step 3, a complex hearing procedure begins. Governments and international industry, trade, and consumer organizations, as well as other committees that may have an interest in the subject, are consulted. Provided a draft standard is suitably endorsed, it will appear on the agenda of the commission at step 5, for discussion and possible endorsement. Once endorsed at step 5 by the commission, the draft goes back to the relevant committee for another round of discussions and consultations. At least two years later, the draft comes back to the CAC for final agreement at step 8. Once agreed upon by the CAC, it will be up to national governments to introduce the standard in national legislation.

The World Trade Organization's (WTO) decision to recognize the Codex standards as the international norm under the Application of Sanitary and Phytosanitary Measures (SPS Agreement) of the General Agreement on Tariffs and Trade (GATT) has amplified the need for horizontal international regulations on contaminants and toxins in food.

The *Codex Alimentarius Commission Procedural Manual* (CAC, 1997b) contains statutes, such as the Internal Regulations for Codex Alimentarius and the Procedure for the Elaboration of Codex Standards, as well as guidelines for governments' acceptance of such standards. The Procedure for Elaboration of Codex Standards contains *inter alia* a detailed description of the stepwise procedure used for the elaboration of such documents. The stepwise procedure ensures that governments are consulted about all draft standards at least twice, that relevant commodity committees overseeing vertical standards are similarly requested to give their opinions, and that the CAC endorses a draft standard at least twice before it is accepted at the highest step 8.

DEVELOPMENT OF THE GSCTF

The Philosophy Paper to the 24th CCFAC 1992

Acting upon the request from the 23rd CCFAC in 1991, the governments of the Netherlands and Denmark prepared the general philosophy paper entitled *Contaminants in Food. Towards a Codex Approach* (Bal & Berg, 1991) for discussion at the 24th CCFAC. This paper stands upon the shoulders of a previous paper by H. P. Mollenhauer (1983), entitled *Contaminants in Food; Approaches and Possible Actions by the Codex Committee on Food Additives,* which served as a guideline for the contaminants work of the CCFAC until this new philosophy paper was presented.

The ideas presented in the Dutch/Danish philosophy paper, elaborated by Dr. Aat Bal of the Ministry of Agriculture, Nature Management and Fisheries of the Netherlands, and the author, became the basis of the GSCTF. The paper discussed the Codex definition of a contaminant and used the origin of the food contamination as a basis for the classification of the contaminants. The contaminants included in present Codex standards and in proposals discussed by the CCFAC were reviewed. This initially led to a discussion of source-directed measures to reduce contamination, and secondly on how to set national and international limits for contaminants in food. Another discussion of when food contaminants require Codex action led to the establishment of criteria, which served as a basis for the CCFAC work as well as to some recommendations to the committee.

For the purpose of the Codex Alimentarius, a contaminant was defined as: "Any substance not intentionally added to food, which is present in such food as a result of the production (including operations carried out in crop husbandry, animal husbandry and veterinary medicine), manufacture, processing, preparation, treatment, packing, packaging, transport or holding of such food or as a result of environmental contamination. The term does not include insect fragments, rodent hairs and other extraneous matter" (CAC, 1997b).

For the work of the CCFAC, it is important that pesticide residues and residues of veterinary drugs are treated in the Codex Committee on Pesticide Residues (CCPR) and in the Codex Committee on Residues of Veterinary Drugs in Foods (CCRVDF), respectively. In some cases, there can be overlap, in particular with the CCPR and chlorinated hydrocarbons. The philosophy paper highlighted that the Codex definition of a contaminant (above) did not include the inherent natural toxins (e.g., the glucosinolates and phycotoxins) since these substances are present in food as a result of the metabolic processes in the organism.

The application of MLs was discussed, as well as the use of guideline levels, prominent in standards at that time.

Use of guideline levels in Codex was defined in connection with radionuclides as: "Guideline Levels are intended for use in regulating foods moving in international trade. When the Guideline Levels are exceeded, governments should decide whether and under what circumstances the food should be distributed within their territory or jurisdiction" (CAC, 1989).

The use of guideline levels and MLs, respectively, will be discussed further in the final section of this chapter.

Various government policies to deal with contaminants in food were discussed. In debating MLs, the paper highlighted the factors of cost versus benefit, quoting an earlier contribution to the committee's work by Beacham (1974), and stated that the MLs should be set as low as reasonably achievable (ALARA).

The paper recommended that a decision on whether national or international action should be taken on a contaminant in a food should be based upon the following criteria:

- The substance is demonstrated to be present in the food at a certain level, which is determined by reliable analysis.
- The substance is of toxicological concern at this level.
- The foodstuff for which action is taken plays a sufficiently important role in the intake of the substance concerned.
- The foodstuff concerned appears in international trade.

An appendix presented a decision-tree approach as an example on how to decide whether a Codex ML should be laid down for a contaminant. Finally, the paper emphasized the importance of source-directed measures to reduce contamination of foodstuffs, in particular for industrial contaminants.

The recommendations presented to the 24th CCFAC were to

- Apply a horizontal approach to all contaminants that can be present in more than one food.
- Review contaminant provisions in existing Codex standards.
- Apply systematically the criteria presented in the paper when making decisions about contaminants.
- Apply source-directed technological measures to prevent contamination and, in the shape of appropriate recommendations, make them part of the Codex system.
- Use dietary recommendations to reduce or eliminate dietary exposure on a national and regional level.
- State that naturally occurring toxins are part of the CCFAC work.
- Develop cooperation between CCFAC and United Nations Environment Programme (UNEP) as well as other similar international programs that aim to reduce pollution and the contamination of food.

The 24th CCFAC in 1992

The 24th CCFAC very clearly marked the shift toward a horizontal approach both for food additives and contaminants, and this was reflected already in the opening speech by the Dutch State Secretary J.D. Gabor (CAC, 1992).

The philosophy paper (Bal & Berg, 1991) was well received and discussed in detail by the CCFAC. The CCFAC agreed with the large majority of the recommendations made in the paper, in particular with the horizontal approach, with the ideas concerning source-directed measures and the criteria proposed for making decisions. Moreover, the committee identified a number of areas that needed further clarification, including the role of good manufacturing practice (GMP) in establishing levels and the meaning and application of guideline levels.

The CCFAC further agreed that the CAC should be requested to revise the CCFAC terms of reference to include naturally occurring toxins because these inherent naturally occurring toxins might need a different approach than the other toxins.

At the conclusion of the discussion, the committee agreed that the delegations of Denmark and the Netherlands would prepare a document detailing the procedures for establishing a general standard for contaminants in food consistent with the above principles for the 25th CCFAC (CAC, 1992).

The 39th Session of the Executive Committee of the CAC discussed these recommendations from the CCFAC and recommended in turn to the CAC to consider incorporating source-directed measures into the Codes of Hygienic Practice. The Executive Committee, moreover, strongly urged the CCFAC to concentrate on the elaboration of MLs for contaminants moving in international trade, giving priority to those contaminants that give the greatest problems in trade, and for which it could be shown that the establishment of levels was required to protect the consumer. With respect to naturally occurring toxins, the Executive Committee expressed concern that levels or limits might cause unnecessary barriers to trade, and that limits should be established only when a health hazard was identified.

The Procedure Paper to the 25th CCFAC in 1993

The Danish and the Dutch delegations took careful note of the positions of both the 24th CCFAC and the Executive Committee when preparing the procedure paper for the 25th CCFAC in 1993. The *Proposed Draft Procedures for the Establishment of a General Codex Standard for Contaminants in Food* (Berg & Kloet, 1992) was presented as the basis for future work. Dr. David Kloet had replaced Aat Bal as representative of the Netherlands and is coauthor of this and all subsequent joint Dutch/Danish papers to the CCFAC on the GSCTF.

The 25th CCFAC discussed this fundamental procedure paper thoroughly and agreed that the two delegations should prepare a proposed draft General Standard for Contaminants in Foods (GSCF) consistent with the paper, particularly section V, which outlined the principles and procedure for establishing the GSCF and the recommendations of the paper (Berg & Kloet, 1992). In particular:

1. The horizontal approach should be used as the basis for the GSCF.
2. The provisions on contaminants in existing Codex commodity standards should be reviewed and revised in accordance with the principles and criteria of the GSCF. Following the revision, such provisions should be replaced by a reference to the GSCF. Work to that effect should be started.
3. The proposed format for the GSCF should be applied.

Comments from the Committee (CAC, 1993d) included:

- The principal aim of the GSCF is consumer protection, whereas priorities on future action would be on preventing barriers to trade.
- Reference to animal feeding stuffs should be included in the scope of the GSCF.
- Action on inherent naturally occurring toxicants would be taken on a case-by-case basis.
- A statement to the effect that the level of a food contaminant should always be ALARA should be included in the beginning of the GSCF.
- MLs should be established wherever possible.

Notes for every contaminant should be prepared according to an agreed format to facilitate the discussion of the committee, and Switzerland accepted to prepare such a note as an example (CAC, 1993b). The question on source-directed measures was solved by Sweden's accepting to draw up a code of practice on such measures, which it did (CAC, 1994b).

The 20th session of the CAC in July 1993 agreed to the above recommendations from the CCFAC.

The First Draft GSCF at the 26th CCFAC in 1994

As agreed, Denmark and the Netherlands prepared the first *Proposed Draft General Standard for Contaminants in Foods* in the autumn of 1993 (CAC, 1993c), and the paper was circulated for comments as part of the preparation of the 26th CCFAC. The GSCF consisted of a Preamble, which presents the scope of the standard; definition of the terms used; the principles and procedures to be applied, which had been amended as agreed; and a description of the format. Annexes comprise:

1. criteria for the establishment of MLs
2. the procedure for risk management decisions in CCFAC regarding contaminants
3. format of the standard
4. numbering system for contaminants to be used in the GSCF
5. food categorization system to be used in the GSCF

An example was included in the paper (CAC, 1993c).

The 26th CCFAC in March 1994 discussed the proposed GSCF at step 3 fully and agreed on some clarification and changes to important details, most importantly:

- to change the title to General Standard for Contaminants and Toxins in Food (henceforth GSCTF)
- to use the term maximum level throughout the GSCTF
- to remove all references to quality matter since the GSCTF addresses public health issues only
- to establish ALARA MLs as a result of good agricultural and manufacturing practice
- to include processed foods in the food categorization system

The CCFAC decided to forward the Preamble of the GSCTF to the Executive Committee for adoption at step 5. Denmark and the Netherlands would then revise the remaining sections and send the revised document for circulation to governments at step 3. The delegation of Switzerland presented a paper with an example on the procedure for inclusion of a contaminant in the GSCTF, and it was agreed that contaminants would be included in the standard on the basis of this example (CAC, 1993b).

The 27th CCFAC 1995

Following the adoption in June 1994 of the Preamble of the GSCTF at step 5 by the Executive Committee, Denmark and the Netherlands elaborated a document for the 27th CCFAC (CAC, 1994c), consisting essentially of the revised set of annexes of the GSCTF.

The 27th CCFAC discussed separately the draft Preamble to the GSCTF at step 7 and the annexes at step 5. With minor additions to the wording, it was agreed to forward the Preamble to the CAC for adoption at step 8 (CAC, 1995e).

The Joint FAO/WHO Consultation on the Application of Risk Analysis to Food Standards Issues had taken place at the WHO headquarters in Geneva shortly before the 27th CCFAC (WHO, 1995), and it was agreed that the authors should harmonize the text of both the Preamble and the annexes of the GSCTF with the recommendations of this important consultation.

The 27th CCFAC agreed after another full discussion that Annexes I, II, and III, duly revised in accordance with the outcome of the debate, should be forwarded to the CAC meeting in July 1995 for adoption at step 5. Annexes IV and V, which did not form an essential part of the standard, but rather served as background documents, would be revised by the authors and circulated again for comments.

The 21st CAC adopted the GSCTF Preamble at step 8 and the Annexes I, II, and III at step 5 in July 1995 (CAC, 1995d).

The 28th CCFAC in 1996

The 28th CCFAC met in Manila, under the chairmanship of Mr. Hans van der Kooi of the Ministry of Agriculture of the Netherlands, and had the draft annexes of the GSCTF on the agenda at step 7. Government comments received were discussed, and the committee stressed that MLs should be established only for contaminants that present a significant risk to public health and a known or expected problem in international trade. The Annexes I, II, and III were again revised by the authors in light of the comments received before and during the session, and the 28th CCFAC agreed that they should be forwarded to the 22nd CAC in 1999 for final adoption at step 8 (CAC, 1996).

The annotated list of contaminants and toxins in Annex IVA and B (CAC, 1996) was considered to be a valuable source of general information; however, the integral incorporation of this annex in the GSCTF was not considered to be necessary. The document would be kept updated, and all delegations were invited to contribute.

It was further agreed that the food categorization system in Annex V should be consistent with the systems already developed for food additives and pesticide residues. Additional classes and product descriptions might, however, be needed in the GSCTF. The Annexes IV and V were forwarded to the Executive Committee at step 5, following revision by the authors in accordance with the CCFAC debate.

The development of position papers on the various contaminants that might eventually be candidates for inclusion in the GSCTF gained momentum in the 28th CCFAC. Already in the 26th CCFAC, Sweden and Denmark had presented a *Discussion Paper on Lead* (CAC, 1993a) as a basis for a future standard, and France had presented a *Position Paper on Cadmium* (CAC, 1994a) to the 27th CCFAC with a similar purpose.

The 28th CCFAC continued work on these contaminants and also received and discussed for the first time a *Position Paper on Aflatoxins* by the United Kingdom (CAC, 1995b) and a *Position Paper on Ochratoxin A* by Sweden (CAC, 1995c). Offers to prepare more position papers for the 29th CCFAC were accepted from the delegations of France and the United Kingdom (patulin); from Australia, Indonesia, and Thailand (tin); and from Denmark (arsenic). The new chair skillfully established priorities for the committee by suspending the collection of information on polychlorinated biphenyls (PCBs), dioxins, polycyclic aromatic hydrocarbons, and hydrogen cyanide, with the understanding that action could be taken at a future meeting if new information became available on health or trade problems. For dioxins and PCBs, the Netherlands had established a discussion paper (CAC, 1995a) that eventually would be updated.

The 29th CCFAC 1997

When the committee met again in 1997 in The Hague, the keynote speaker, Director-General of the Dutch Ministry of Agriculture, J. F. de Leeuw, emphasized the increasing importance of the committee's work in the context of implementing the World Trade Agreements.

The 43rd Session of the Executive Committee had already in July 1996 adopted Annexes IV and V to the GSCTF at step 5, which had been circulated for government comments at step 6. At this stage, only a few comments were received (CAC, 1995a), and after a short debate the committee decided to forward the introduction of Annex IV and the whole of Annex V to the CAC for adoption at step 8 (CAC, 1997e).

The work on position papers on lead, cadmium, tin, and arsenic as well as on the aflatoxins, ochratoxin A, and patulin progressed, and an offer from Norway to produce a position paper on zearalenone was accepted by the CCFAC.

The 22nd Session of the Codex Alimentarius Commission in July 1997 adopted all five annexes of the GSCTF at step 8, completing the framework of the general standard (CAC, 1997d).

The 30th CCFAC 1998

Having the GSCTF adopted by the CAC provided the CCFAC with a complete framework, but nothing more than that. Moreover, individual efforts by single delegations or small teams provided the committee with revised position papers (or equivalent) on most, or perhaps all, relevant contaminants. Other delegations, however, were more reluctant to accept the content of the papers, and fairly few proposals of MLs for individual contaminants were advanced in the stepwise procedure.

A discussion of a paper produced by the United Kingdom delegation on *Methodology and Principles for Exposure Assessment in the Codex General Standard for Contaminants and Toxins in Food* (CAC, 1997c) took place. The paper was developed with assistance of the Netherlands and Denmark. In the debate, it was emphasized that the Joint FAO/WHO Expert Consultation on Food Consumption and Exposure Assessment of Chemicals in February 1997 (WHO, 1997) should be taken into account, and that it remains important to address major food groups as well as individual foods of importance. Regional dietary differences should be considered, too. The methodology should give preference to the ALARA principle and should not be based upon the highest levels of contaminants observed. The committee decided to develop the work further, and the United Kingdom delegation agreed to take responsibility for that.

At the end of the session, Chairman Edwin Hecker of the Netherlands concluded that in spite of the great amount of work completed over the last decade, there were still a large number of contaminants on the CCFAC agenda (CAC, 1997f). He addressed this problem by recommending to the CCFAC to install an ad hoc Working Group (WG) on Contaminants, and the committee accepted his proposal. Denmark was appointed chair of the ad hoc WG, with Brazil as cochair. The WG met for the first time immediately before the 31st CCFAC in 1999 (CAC, 1998).

The goals of the ad hoc WG are to achieve progress in the further development of the GSCTF and make recommendations to the plenary session of the CCFAC on the issue of contaminants.

The terms of reference of the ad hoc WG include to propose refinements of the GSCTF and review proposals to the Joint FAO/WHO Expert Committee on Food Additives (JECFA) for contaminants to be evaluated.

The 31st CCFAC 1999

Introduction of the ad hoc WG on Contaminants and Toxins in Food did indeed accelerate the development concerning establishment of MLs for contaminants in food. The combination of a preliminary, mainly technical discussion in the WG and, if necessary, a full debate in the plenary session, enabled the committee to cover more ground than had been the case ever before.

Position papers on zearalenone by Norway, on patulin by France, and on arsenic by Denmark were accepted, and the delegations were asked to finalize them. For zearalenone and also for ochratoxin A, a draft Code of Practice for the reduction of contamination will be developed. For arsenic, future work will have to await that routine methodology for distinction between the generally more toxic inorganic arsenic species and the less toxic organic compounds to become available.

For ochratoxin A, a draft ML of 5 µg/kg in cereals and cereal products will be circulated for comments at step 3, with an appropriate sampling plan, while a JECFA evaluation is awaited. The draft ML for patulin in apple juice and apple juice ingredients in other beverages of 50 µg/l was sent to CAC at step 5, and the United States offered a position paper on fumonisins for the 32nd Session (CAC, 1999a).

The proposal for MLs for lead remained at step 7 after a thorough discussion, since JECFA is to evaluate lead in relation to children in particular, in 1999, whereas a draft series of MLs for cadmium in a number of foods was accepted for circulation at step 3, with a ML for cereals coming from the Codex Committee on Cereals, Pulses and Legumes (CCCPL) already at step 6. The MLs for tin were advanced to CAC for adoption at step 5, and it was agreed to ask JECFA to reassess the acute toxicity of tin as a matter of priority (CAC, 1999b).

Finally, it was agreed to review the Codex commodity standards, and draft standards, for their MLs for contaminants, in order to remove remaining references to zinc, iron or selenium as contaminants—they may remain as quality parameters if deemed appropriate by the relevant commodity committee—and to identify provisions for lead, cadmium, tin, or arsenic for revision in the GSCTF context.

The revised U.K. paper on Methodology and Principles for Exposure Assessment in the GSCTF (CAC, 1999b) was again discussed, and it was agreed that the intent and objectives would be carefully considered by a drafting group led by the United Kingdom, and that a redrafted document in the form of an annex to the GSCTF would be developed by the drafting group.

In accordance with the results, it was decided to reinstall the ad hoc WC, with the author as chair, before the 32nd Session.

As it will follow from the above description of the development of the GSCTF, the status as of 1999 is that the text of the standard is fully accepted by the CAC, and it is becoming increasingly clear which contaminants will have MLs. It remains, however, to be finally agreed on for which contaminants there are to be MLs, and what these MLs in various food groups or individual foods will be.

SCOPE, PURPOSE, AND PRINCIPLES OF THE GENERAL STANDARD ON CONTAMINANTS AND TOXINS IN FOOD

The GSCTF contains the main principles and procedures that are used by the Codex Alimentarius in dealing with contaminants in foods and feeds. In due course when completed, it will list the MLs of contaminants and natural toxicants in foods and feeds to be applied to commodities moving in international trade. This section highlights some of the more important sections of the GSCTF.

Foods and feeds can become contaminated for various reasons and by many processes. Generally, contamination has a negative impact on food quality and may imply a risk to human or animal health. Contaminant levels in food should follow the ALARA principle. The following actions may serve to prevent or reduce contamination of food:

- preventing food contamination at the source (e.g., by reducing environmental pollution)
- applying appropriate food technology in food production, handling, storage, processing, and packaging
- applying measures aimed at decontamination of contaminated food and measures to prevent contaminated food and feed from being marketed for consumption

A general *Draft Code of Practice for Source-Directed Measures To Reduce Contamination of Foodstuffs* was developed by the Swedish delegation and presented to the 30th CCFAC (CAC, 1998), based upon a previous draft (CAC, 1994b). The CCFAC agreed to circulate this document for comments at step 3.

Specific codes of practice may be, and should certainly be, developed in cases when such codes of practice might serve to improve food quality. The *Code of Practice for the Reduction of Aflatoxins in Raw Materials and Supplementary Feedingstuffs for Milk Producing Animals* (CAC, 1997a) is an example of a code of practice.

The degree of contamination of food and the effect of action to reduce food contamination shall be assessed by monitoring survey programs and research programs, with emphasis on quality assurance of the data provided. When there are indications that health hazards may be involved with the consumption of foods that are contaminated, a risk assessment will be made. Attention will be paid to the vocabulary used for this purpose since the definitions of the terms for Codex Alimentarius purposes are being developed (WHO, 1995; CAC, 1997f; and FAO, 1997). When health concerns can be substantiated, a risk management policy must be applied, based upon the thorough evaluation of the situation. It may be necessary to establish MLs or other measures governing the contamination of foods for sale or, in special cases, to give dietary recommendations when other measures are not adequate to exclude the possibility of hazards to health of the consumer.

National measures regarding food contamination should, when inevitable, be established so that the creation of unnecessary barriers to international trade in food commodities is avoided. The purpose of the GSCTF is to provide guidance about possible approaches to food contamination problems and to promote international harmonization through recommendations that may help to avoid barriers to trade caused by contaminants.

MLs will only be set for those foods in which the contaminant may be found in amounts that are significant for the total exposure of the consumer. They shall be set in such a way that the consumer is adequately protected. At the same time, the technological possibilities to comply with the MLs will be taken into account. The principle of GMP, good veterinary practice (GVP), and good agricultural practice (GAP) will be used. MLs will be based on sound scientific principles, thereby leading to worldwide acceptable MLs so that international trade in food is facilitated.

Specific Criteria

The following specific criteria will be considered when recommendations and decisions are made in connection with the GSCTF (e.g., when the notes on each contaminant in Annex IV of the GSCTF are developed or when position papers on individual contaminants are written).

Toxicological Information

Criteria for toxicological information are as follows:

- identification of the toxic substance(s)
- metabolism by humans and animals, as appropriate
- toxicokinetics and toxicodynamics
- information about acute and long-term toxicity and other relevant toxicity
- integrated toxicological expert advice regarding the acceptability and safety of intake levels of contaminants, including information on any population groups that are specially vulnerable

A recommendation from JECFA regarding the maximum tolerable intake, based upon a full evaluation of an adequate toxicological database, should be the main basis for decisions from CCFAC. In urgent cases, it may be necessary to rely on less developed evaluations from JECFA or on toxicological expert advice from other international or national bodies.

Analytical Data

Criteria for analytic data include:

- validated qualitative and quantitative data on representative samples
- appropriate sampling procedures

Information on the methods of sampling and analysis applied as well as on the validation of the obtained results is desirable. The portion of the commodity that was analyzed should be clearly given and should preferably be equivalent to the definition of the commodity with respect to a regulation. A statement on the representative nature of the samples should be added. Special attention should be paid to sampling in the case of contaminants that may be unequally distributed in the products, such as aflatoxins in peanuts or figs.

Intake Data

Criteria for intake data are as follows:

- presence in foods of dietary significance for the contaminant intake
- presence in foods that are widely consumed
- food intake data for average and most exposed consumer groups

- results from total diet studies
- calculated contaminant intake data from food consumption models
- data on intake by susceptible groups

It is desirable to have information about the contaminant concentrations in those foods or food groups that are responsible for at least half and preferably 80% or more of the total dietary intake of the contaminant, both for average and for high consumers.

WHO guidelines for the study of dietary intake of chemical contaminants should be taken into consideration when intake studies are used for the assessment of exposure of the consumer (WHO, 1997). It is important to be aware of the type of the dietary study in order to use it correctly for exposure calculations. Results on the effect of preparation and cooking on the concentration levels of a contaminant (reduction factors) may also be useful.

Fair Trade Considerations

Fair trade considerations are:

- existing or potential problems in international trade
- commodities concerned moving in international trade
- information about national regulations, in particular on the data and considerations on which these national regulations are based

Technological Considerations

Technological considerations include information about contamination processes, technological possibilities, production and manufacturing practices, and economic aspects related to contaminant ML management and control.

The most efficient ways to control food contamination are often technological ways and means to reduce the contamination of the raw product or to reduce the concentration of the contaminant in the final food. Source-directed measures may be beyond the scope of the CCFAC, but they may nevertheless often be the ones that efficiently reduce or remove a food contamination problem. Internationally agreed GAP and GMP will serve to make it possible to apply the ALARA principle for the benefit of the consumer as well as for the reputation of international trade in foods when instituting MLs for contaminants in foods.

Risk Assessment and Risk Management Considerations

Risk assessment and risk management considerations include:

- risk assessment
- risk management options and considerations
- consideration of possible MLs in foods based upon the criteria mentioned above
- consideration of alternative solutions

Risk assessment in food science is defined as the scientific evaluation of the probability of occurrence of known or potential adverse health effects resulting from human exposure to food-borne hazards. The process consists of hazard identification, hazard characterization, exposure assessment, and risk characterization (WHO, 1987, and CAC, 1997a).

Risk management is defined as the process of weighing policy alternatives in the light of risk assessment and, if required, to select and implement appropriate control options, including the establishment and enforcement of MLs for contaminants in foods. It is based upon adequate risk assessment and on information about policy options and strategies to deal with contamination problems. Risk management includes risk communication. Risk management involves, in a consistent way, to decide what is acceptable and what is not in a given situation, and to decide what should be done to achieve sufficient protection of public health as well as control of the contamination.

There may be other criteria than those listed above that are relevant when elaborating a position paper for a particular contaminant or, more likely, there may not be detailed information available corresponding to all the specific criteria above. Neither situation should lead to the conclusion that nothing further can be done.

It is important to bear in mind that new data will become available, or perhaps that a more thorough JECFA assessment of data will be provided. Meanwhile, the CCFAC should act on the data and assessments available so that the GSCTF can be completed for the benefit of consumers as well as for the international trade in food. When in the future, better data and assessments come to light, MLs and the standards can be improved.

THE CONTENT OF THE GENERAL STANDARD—STATUS AND FUTURE

The GSCTF is at present (1999) only partially complete. Molière might have described the language of the GSCTF as beautiful—a recipe for a lovely soup where the ingredients were all there, but the soup had yet to be cooked.

The identification of contaminants that pose problems to consumers' health or to international trade in foods continues. Work continues to provide the CCFAC with an appropriate basis to make risk management decisions in the form of standards, or MLs, for those contaminants in relevant types of food. These MLs can then be integrated into the GSCTF.

The format of the GSCTF is presented in Annex III of the standard. Annex V describes the food categorization system to be used, which is essentially the one developed in the framework of the CCPR. This system is especially elaborated to accommodate primary agricultural commodities, and needs further extension regarding processed products. Where necessary, new subgroup codes or commodity codes are to be introduced according to an agreed principle used in other Codex standards.

Work is in progress to identify the contaminants for which the CCFAC shall elaborate MLs, and a list will most probably comprise the following elements: lead, cadmium, tin, and eventually arsenic, as well as the mycotoxins (aflatoxins, ochratoxin A, and patulin), and possibly fumonisins. The list may also include one or two more substances or groups of substances, such as PCBs or dioxins. The time for a decision to focus the attention of the committee on a limited number of substances, or groups of substances, is near. The CCFAC should help as much as possible to finish the position papers and the draft standards currently on the table until final adoption by the CAC.

Other contaminants remain of interest, but will probably have to get second priority for a while. In the sections dealing with contaminants in existing Codex commodity standards, other substances than those listed above are also mentioned, such as iron, copper, and zinc. Limits to their presence in foods are to be considered as quality parameters for the food products, rather than as measures for the contamination of the products. A decision not to include limits to these parameters in the future, at least not as contaminant provisions, was taken by the 31st CCFAC.

For the time being, there are no naturally occurring toxicants on the contaminant part of the agenda of the CCFAC. This does not mean that the CCFAC debate about how to deal with such substances has been superfluous. A series of naturally occurring phytotoxins and other inherent naturally occurring toxicants are in section 7 of Annexes IVA and B. Whenever such substances give rise to public health or international trade problems, the mechanism to deal with such problems under the auspices of the GSCTF is in place.

During the development of the GSCTF, there have been several CCFAC debates concerning the use of the term *guideline levels* (or *limits*). Time, or rather the introduction of the Agreement on the Application of Sanitary and Phytosanitary Measures of the WTO, seems to have rendered any such discussions superfluous in the future since any Codex Alimentarius limit or level is to be considered a binding standard in trade disputes between member state signatories to the agreement.

REFERENCES

Bal, A. & Berg, T. (1991). *Contaminants in Food. Towards a Codex Approach.* CX/FAC 92/10. Rome: Codex Alimentarius Commission.

Beacham, L. M. (1974). *The Scope and Magnitude of Work for the Setting up of International Standards for Contaminants and Pollutants in Foods.* ALINORM 76/29. Rome: Codex Alimentarius Commission.

Berg, T. & Kloet, D. (1992). *Proposed Draft Codex Procedure for the Establishment of a General Codex Standard for Contaminants in Food.* CX/FAC 93/11. Rome: Codex Alimentarius Commission.

Codex Alimentarius Commission. (1989). *Codex Alimentarius Commission,* Vol. XVII—ed. I, suppl. 1. CAC/GL 5. Rome: Codex Alimentarius Commission.

Codex Alimentarius Commission. (1991). *Report of the 23rd Session of the Codex Committee on Food Additives and Contaminants.* ALINORM 91/12A, paras. 22–28. Rome: Codex Alimentarius Commission.

Codex Alimentarius Commission. (1992). *Report of the 24th Session of the Codex Committee on Food Additives and Contaminants.* ALINORM 93/12, paras. 2–4 and 64–78. Rome: Codex Alimentarius Commission.

Codex Alimentarius Commission. (1993a). *Discussion Paper on Lead*. CX/FAC 94/20. Rome: Codex Alimentarius Commission.

Codex Alimentarius Commission. (1993b). *Procedure for the Inclusion of a Contaminant in the Codex General Standard for Contaminants in Foods*. CX/FAC 94/13. Rome: Codex Alimentarius Commission.

Codex Alimentarius Commission. (1993c). *Proposed Draft General Standard for Contaminants in Foods*. CX/FAC 94/12. Rome: Codex Alimentarius Commission.

Codex Alimentarius Commission. (1993d). *Report of the 25th Session of the Codex Committee on Food Additives and Contaminants*. ALINORM 93/12 A, paras. 106–115. Rome: Codex Alimentarius Commission.

Codex Alimentarius Commission. (1994a). *Position Paper on Cadmium*. CX/FAC 95/19. Rome: Codex Alimentarius Commission.

Codex Alimentarius Commission. (1994b). *Proposed Draft Code of Practice for Source-Directed Measures To Reduce Contamination of Food with Chemicals*. CX/FAC 94/19. Rome: Codex Alimentarius Commission.

Codex Alimentarius Commission. (1994c). *Report of the 26th Session of the Codex Committee on Food Additives and Contaminants*. ALINORM 95/12, paras. 92–110. Rome: Codex Alimentarius Commission.

Codex Alimentarius Commission. (1995a). *Discussion Paper on PCBs and Dioxins*. CX/FAC 96/25. Rome: Codex Alimentarius Commission.

Codex Alimentarius Commission. (1995b). *Position Paper on Aflatoxins*. CX/FAC 96/18. 1995, Rome: Codex Alimentarius Commission.

Codex Alimentarius Commission. (1995c). *Position Paper on Ochratoxin A*. CX/FAC 96/21. Rome: Codex Alimentarius Commission.

Codex Alimentarius Commission. (1995d). *Report of the 21st Session*. Rome: Codex Alimentarius Commission.

Codex Alimentarius Commission. (1995e). *Report of the 27th Session of the Codex Committee on Food Additives and Contaminants*. ALINORM 95/12 A, paras. 93–104. Rome: Codex Alimentarius Commission.

Codex Alimentarius Commission. (1996). *Report of the 28th Session of the Codex Committee on Food Additives and Contaminants*. paras. 63–70. Rome: Codex Alimentarius Commission.

Codex Alimentarius Commission. (1997a). *Code of Practice for the Reduction of Aflatoxins in Raw Materials and Supplementary Feedingstuffs for Milk Producing Animals*. Rome: Codex Alimentarius Commission.

Codex Alimentarius Commission. (1997b). *Codex Alimentarius Procedural Manual*, 10th ed. Rome: Codex Alimentarius Commission.

Codex Alimentarius Commission. (1997c). *Methodology and Principles for Exposure Assessment in the Codex General Standard for Contaminants and Toxins in Food*. CX/FAC 98/13. Rome: Codex Alimentarius Commission.

Codex Alimentarius Commission. (1997d). *Report of the 22nd Session*. Rome: Codex Alimentarius Commission.

Codex Alimentarius Commission. (1997e). *Report of the 29th Session of the Codex Committee on Food Additives and Contaminants*. ALINORM 97/12A, paras. 51–53. Rome: Codex Alimentarius Commission.

Codex Alimentarius Commission. (1997f). *Report of the 30th Session of the Codex Committee on Food Additives and Contaminants.* ALINORM 99/12. Rome: Codex Alimentarius Commission.

Codex Alimentarius Commission. (1998). *Draft Code of Practice for Source-Directed Measures To Reduce Contamination of Foodstuffs.* CX/FAC 98/20. Rome: Codex Alimentarius Commission.

Codex Alimentarius Commission. (1999a). *Report of the 31st Session of the Codex Committee on Food Additives and Contaminants.* ALINORM 99/12A. Rome: Codex Alimentarius Commission.

Codex Alimentarius Commission. (1999b). *Methodology and Principles for Exposure Assessment in the GSCTF.* CX/FAC 99/13. Rome: Codex Alimentarius Commission.

Denner, W. H. B. (1989). *Future Activities of the Committee in Regard to the Establishment and Regular Review of Provisions Related to Food Additives in Codex Standards and Possible Mechanisms for the Establishment of General Provisions for the Use of Food Additives in Non-standardized Foods.* CX/FAC 89/16. Rome: Codex Alimentarius Commission.

Food and Agriculture Organization. (1997). *Application of Risk Management to Food Safety Matters.* Joint FAO/WHO Expert Consultation, Rome, January 27–31, 1997. FAO Food and Nutrition Paper 65. Rome: Food and Agriculture Organization.

Mollenhauer, H. P. (1983). *Contaminants in Food; Approaches and Possible Actions by the Codex Committee on Food Additives.* CX/FA 83/18. Rome: Codex Alimentarius Commission.

World Health Organization. (1987). *Guidelines for the Study of Dietary Intakes of Chemical Contaminants.* WHO Offset Publication No. 87. Geneva, Switzerland: World Health Organization.

World Health Organization. (1995). *Application of Risk Analysis to Food Standard Issues. Report of the Joint FAO/WHO Expert Consultation.* Geneva, Switzerland, March 13–17, 1995. Geneva, Switzerland: World Health Organization.

World Health Organization. (1997). *Food Consumption and Exposure Assessment of Chemicals. Report of a FAO/WHO Consultation.* Geneva, Switzerland, February 10–14, 1997. Geneva, Switzerland: World Health Organization.

CHAPTER 13

Codex Standards for Pesticide Residues

Gerald G. Moy and John R. Wessel

INTRODUCTION

The use of pesticides[1] in agriculture provides numerous benefits to the farmer and consumer alike. Pests in the form of insects, weeds, molds, and plant diseases pose a threat to the production of a plentiful, wholesome, and economic food supply. Indeed, without the use of pesticides to mitigate pests, there would likely be serious crop shortfalls throughout the world. Surviving crops would probably be of inferior quality, prone to spoilage, and perhaps even unsafe for consumption. Notwithstanding these benefits, unregulated use of pesticides poses unacceptable risks to the health of consumers and the environment.

To ensure that the public is protected from such risks, countries have adopted laws governing pesticide use that include provisions for residues of pesticides on food. Most national laws require that a pesticide must first be registered (i.e., approved or licensed) for use based upon a demonstration that the pesticide is effective for its intended use and that the use will not adversely affect public health or the environment. The culmination of the registration process is the approval of the pesticide's proposed label directions, which for agricultural use, specifies, among other things, target pest(s) or purpose of use, the identity and percent of active pesticide ingredient, application rate, frequency of application, and preharvest interval (i.e., number of days before a crop can be harvested after a pesticide's application). Label directions form part of what is referred to as good agricultural practice (GAP) in the use of a pesticide (FAO/WHO, 1997a).

To protect public health and ensure proper use of pesticides in agriculture, national laws require the establishment of a tolerance or maximum residue limit (MRL), which defines the maximum amount of a pesticide residue that may legally

---

[1] In the context of this chapter, the term *pesticides* includes substances intended for use as insecticides, plant-growth regulators, defoliants, desiccants, fruit-thinning agents, or sprouting inhibitors as well as substances applied after harvest to prevent deterioration (FAO/WHO, 1997a).

remain in or on a food commodity[2] after the pesticide is used according to its GAP. Thus, when a pesticide is applied in accordance with its labeled directions, the MRL should not be exceeded. Establishment of the MRL also requires a demonstration that residues of the pesticide on food do not pose a risk to public health. As such, an MRL serves as an enforcement tool by providing a means of ensuring that both the GAP is followed and public health is protected.

NEED FOR INTERNATIONAL STANDARDS FOR PESTICIDE RESIDUES

While national pesticide laws do not vary greatly among countries, administration of those laws may be quite different (FAO/WHO, 1982). In addition, climatic conditions, endemic pest problems, and agricultural practices cannot only affect the choice of pesticide, but also its application rate, pattern of use, and resulting MRLs. Differences in pesticide use and MRLs may become a trade problem when pesticide-treated food commodities are shipped in international trade to countries having national regulations that conflict with those of the exporting country. Typically, MRLs apply to both domestically produced and imported food, but are established based on the domestic GAP without consideration of GAPs of foreign countries. As a result, imported food commodities may contain pesticide residues that do not comply with an importing country's MRLs and may be subject to regulatory measures, such as rejection or detention.

One solution to these trade problems is for an exporting country to apply only pesticides registered in the importing country in such a manner that residue levels meet the importing country's MRLs. Another possible solution is for the importing country to establish MRLs that accommodate the different pesticide needs of the exporting country. For various reasons, however, these bilateral solutions have been difficult to achieve in practice. Bilateral arrangements also tend to neglect consideration of other countries' requirements. For example, even if a food export can be produced to meet the pesticide limits of one country, it may not comply with the requirements of other importing countries. Therefore, national MRLs can create nontariff barriers to international trade that can affect all countries exporting and importing food. The best solution to the problem is for all countries to use MRLs for imported food that accommodate the GAPs of exporting countries. Such international MRLs would serve to protect the health of consumers while promoting fair trade in food.

ROLE OF INTERNATIONAL ORGANIZATIONS AND THE CODEX ALIMENTARIUS COMMISSION

International organizations, especially the Food and Agriculture Organization (FAO) of the United Nations in Rome and World Health Organization (WHO) in

---

[2]Pesticide residues can occur either in or on a food crop or both. For purposes of this chapter, the expression *on food* will be used to convey each type of occurrence.

Geneva, have important roles to play in the harmonization of health and safety requirements for food. For example, the WHO Constitution of 1949 states that one of the functions of WHO is "to develop, establish and promote international standards with respect to food . . ." (WHO, 1982). In this regard, early concern for the safety of pesticides led to a WHO monograph, *Toxic Hazards of Certain Pesticides*, which was published in 1953 (WHO, 1953). Recognizing the growing importance of international trade in food, the Codex Alimentarius Commission (CAC) was formed in 1962 under the joint sponsorship of FAO and WHO. Convened under the auspices of the Joint FAO/WHO Food Standards Programme, the CAC aims to develop and secure international agreement on food standards, guidelines, and other recommendations related to food that are protective of public health and ensure fair practices in international trade (FAO/WHO, 1997a). CAC, an intergovernmental body that now has over 150 member countries, conducts its work through subsidiary bodies, including general subject committees, specific commodity committees, and regional coordinating committees. The Codex Committee on Pesticide Residues (CCPR) was established by the CAC with responsibility for the development of Codex standards for pesticide residues on food. The CCPR, which is hosted by the government of the Netherlands, receives scientific and technical advice from the Joint FAO/WHO Meeting on Pesticide Residues (JMPR). Since its first meeting in 1966, the CCPR has provided a forum for achieving agreement among member countries on international MRLs for pesticide residues on food[3] that ensure a safe, single standard covering GAPs of countries in their use of pesticides. An overview of the organization of pesticide residue risk management within the framework of the CAC is given in Figure 13–1.

Use of Codex MRLs by member countries is intended to preempt divergent national MRLs from constituting nontariff barriers to international trade in food. The development of Codex MRLs follows well-established procedures that provide member countries with ample opportunity for review and comment (FAO/WHO, 1997a).

SELECTION OF PESTICIDES FOR ELABORATION OF CODEX MRLs

To enter the Codex step procedure (i.e., step 1), a pesticide must first be placed on the CCPR priority list for MRL development. The criteria for the selection process are: (1) the pesticide is available for use as a commercial product, (2) its use gives rise to residues on a food commodity moving in international trade, (3) the presence of the residues may be a matter of public health concern, and (4) it has the potential to create problems in international trade (FAO, 1997). In addition to these criteria, there must be a commitment that all relevant data will be made available for evaluation.

Based on proposals made by national governments, the CCPR reviews and updates the priority list during its annual sessions. The priority list is forwarded to FAO and WHO, which serve as the Secretariat for JMPR (discussed below). Proposals

---

[3]Codex MRLs extend to food commodities of both plant and animal origin and to animal feedstuffs.

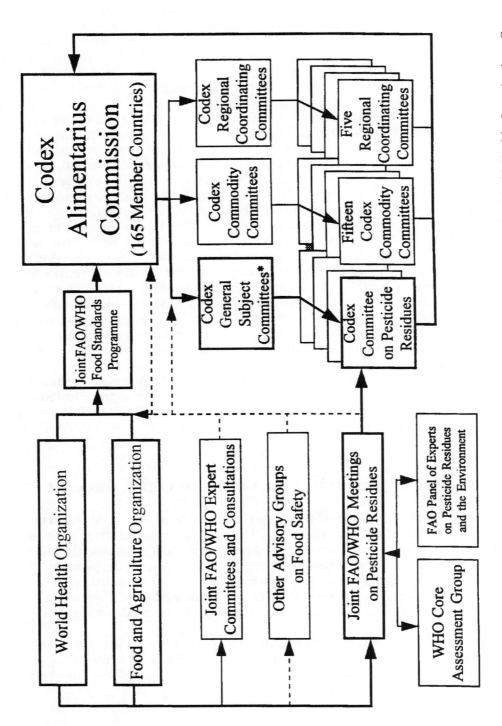

Figure 13–1 Organization of Codex Pesticide Residue Risk Management. Courtesy of the World Health Organization, Geneva, Switzerland.

*Includes nine Codex general subject committees.

for additional MRLs for pesticides that are already in the Codex procedure can be made by any interested party directly to the JMPR Secretariat provided a commitment is made to submit relevant data.

JOINT FAO/WHO MEETING ON PESTICIDE RESIDUES

The JMPR consists of a jointly convened meeting of the FAO Panel of Experts on Pesticide Residues and the Environment and the WHO Core Assessment Group on Pesticides. Begun in 1963, the JMPR has been held annually for the purpose of evaluating pesticides in terms of their occurrence as residues on food and their potential effects on the health of consumers. During this time, JMPR has evaluated approximately 220 pesticides (many of these more than once) and has proposed over 3,000 MRLs. Appointed by FAO and WHO, members of the JMPR are experts in the field of pesticides who serve in their professional capacities as individual scientists and not as representatives of their governments or the institutions in which they are employed. Scientists from the pesticide and food industries do not serve on the JMPR.

The JMPR serves as an independent scientific advisory body for the CCPR as well as for FAO and WHO member countries. Once a pesticide is on the CCPR priority list, the JMPR will evaluate submitted data concerning the pesticide's safety and proper use. This may include proposals for draft MRLs, which constitutes step 2 of the Codex step procedure. The sources of these data are the open, scientific literature and manufacturers of the pesticide. The data are required to be submitted in advance of a JMPR session to permit their review by FAO- and WHO-appointed temporary advisors. These scientists conduct the preliminary evaluations and present their findings to the experts at the JMPR session.[4] During the JMPR, the hazard characterization performed by the WHO Core Assessment Group may result in the establishment of an acceptable daily intake[5] (ADI) for the pesticide. When appropriate, an acute reference dose[6] (acute RfD) for the pesticide may also be established. The principles used by JMPR for the toxicological assessment of pesticide residues were summarized in 1990 (WHO, 1990). At the same time, the FAO panel may propose maximum levels for residues (i.e., proposed draft MRLs) for the pesticide on commodities based on available information on GAP.

At the conclusion of each JMPR, a report is prepared describing the general topics discussed and summarizing the data and recommendations for each pesticide evaluated, including ADIs, MRLs, and predictions of dietary intake. The meeting

---

[4]The use of temporary advisors during the year prior to a JMPR is necessary because the FAO- and WHO-appointed experts, along with the temporary advisors, meet as a group only for one or two weeks.

[5]An ADI is the amount of a chemical, expressed on a body weight basis, that may be ingested over an entire lifetime without appreciable risk to the health of the consumer based on all the known facts at the time of the evaluation.

[6]An acute RfD of a chemical is the amount of a substance in food, expressed on a body weight basis, that can be ingested over a short period of time, usually during one meal or one day, without appreciable health risk to the consumer based on all the known facts at the time of the evaluation.

report is supplemented by two other publications containing detailed information on the evaluations of the JMPR. Published by FAO under its FAO Plant Production and Protection Series, *Pesticide Residues in Food: Part I—Residues* provides information on agricultural and pest control practices, chemistry and metabolism, supervised field trial studies,[7] fate of the pesticide residue in the field and during storage and processing of food, residue analytical methodology, and other relevant information. Published by WHO under the auspices of the International Programme on Chemical Safety (IPCS), *Pesticide Residues in Food: Part II—Toxicology* provides information on physical/chemical properties, toxicity studies in animals and humans, metabolism and estimation of toxicological endpoints, and ADIs.

In the future, it is envisioned that a joint meeting on pesticides will consolidate international activities on pesticide assessment and will include, in addition to the functions already performed by the JMPR, consideration of issues related to public and occupational health and the environment (WHO, 1996).

Proprietary Data

As noted above, one source of data for evaluation by the JMPR is the manufacturer of the pesticide. Data from this source usually are unpublished and are considered proprietary by the manufacturer. This ownership stems from the manufacturer's sponsoring the development of toxicological and other scientific data required to support registration and MRLs in a country. The owner of these data may wish to limit the advantages competitors could gain through the use of these data without sharing the commercial risks and economic costs associated with its development.

Both FAO and WHO have supported the need to protect the proprietary rights of manufacturers who submit pesticide data. IPCS, on behalf of WHO, has adopted operating procedures to safeguard unpublished proprietary data received for JMPR evaluations from unauthorized disclosure. In addition, JMPR reports and monographs include a statement that cautions registration authorities to not grant pesticide registrations based on the JMPR evaluations, which rely on proprietary data, unless authorization for the use of such data is received from the owner of such data or from a second party that has obtained permission for this purpose.

SAFETY OF PESTICIDE RESIDUES

Because consumers worldwide expect a safe food supply, the acceptability of proposed draft MRLs from a consumer safety point of view is a critical consideration for the JMPR. The safety of residues resulting from a pesticide's use in food production is evaluated by a risk assessment that compares potential exposure to residues of the pesticide with its established safe level (e.g., ADI) (WHO, 1997). In

---

[7] A supervised field trial is an experimental study that generates residue data on a food crop treated with a pesticide under proposed maximum conditions of use. The results of a supervised field trial provide a range of residue values that reflect GAP and are used to estimate an appropriate level for an MRL. These data are also used to determine supervised trials median residue (STMR) levels for dietary intake purposes.

general, exposure is determined by knowing the amount of food consumed and the concentration of pesticide residues on that food.

For hazards that are posed by chronic (long-term) exposure to residues, the exposure assessment is based on the average daily dietary intake of a food commodity multiplied by the residue level expected on that commodity. The total dietary exposure is calculated by summing all the contributions from individual food commodities expected to contain the pesticide in question. Since 1989, draft MRLs proposed by the JMPR have been subject to an international risk assessment performed by the Global Environmental Monitoring System/Food Contamination Monitoring and Assessment Programme (GEMS/Food) of WHO in close cooperation with the FAO panel. For screening purposes, GEMS/Food calculates a theoretical maximum daily intake (TMDI) based on proposed and existing Codex MRLs for each of five regional diets. These GEMS/Food regional diets were derived from FAO food balance sheets for selected countries and include African, Far Eastern, Latin American, Middle Eastern, and European-type dietary patterns.[8] The TMDI is calculated by multiplying the proposed draft MRL or existing Codex MRL (if no draft MRL is proposed for the commodity) by the estimated per capita regional consumption for the food commodity and then summing the products of all MRL/commodity combinations as shown in the following equation:

$$TMDI = \sum MRL_i \times F_i$$

where
MRL_i = Proposed draft MRL or existing MRL for a given food commodity
F_i = Per capita average daily regional consumption of that food commodity

The TMDI is compared to the ADI for the pesticide (calculated for a 60-kg person) and is expressed as a percent of the ADI. For the Far Eastern diet, a body weight of 55 kg may sometimes be used to improve the estimate. More precise average body weights will be taken into consideration during the development of future regional diets. While the use of MRLs in calculating the TMDI greatly overestimates residues, TMDI estimates for most pesticides have been below their corresponding ADIs (i.e., the TMDIs were less than 100%). Consequently, the proposed MRLs for these pesticides were considered safe from a public health perspective. However, for certain pesticides, more refined exposure assessments have proven to be necessary.

Since 1996, GEMS/Food has been using the international estimated daily intake (IEDI) to better estimate exposure to pesticide residues. The approach, which was developed in collaboration with the CCPR, relies on median residue levels determined from supervised field trials data to better estimate likely residues on food (WHO, 1997). Unlike MRLs, which are established for compliance purposes, the median residue value from supervised trials (referred to as STMR) is derived specifically for dietary purposes. An STMR level provides a more realistic indication of the actual residue level than an MRL level for a given food commodity. An STMR

---

[8] Copies of the GEMS/Food regional diets are available through the WHO Web Site: http://www.who.ch/ or by writing to the WHO Programme of Food Safety, CH-1211. Geneva, Switzerland.

is defined to include all residues of toxicological concern, such as conversion products, metabolites, reaction products, and impurities. Certain factors that may influence residue levels before the food is consumed are also considered when data are available. These factors include residue levels in edible portions of food commodities and effects of processing and cooking on levels of pesticide residues. The IEDI is calculated by the following equation:

$$\text{IEDI} = \sum \text{STMR level}_i \times E_i \times P_i \times F_i$$

where
- STMR level_i = STMR level for a given food commodity
- E_i = Edible portion factor for that food commodity
- P_i = Processing factor for that food commodity
- F_i = Per capita average daily regional consumption of that food commodity

JMPR-proposed draft MRLs still having dietary intake concerns after all available factors have been taken into account (i.e., the IEDI exceeds the pesticide's ADI) are referred to the CCPR for consideration of possible risk management measures that may mitigate the intake concerns (FAO/WHO, 1997b).

Some pesticides can also produce acute health effects from short-term exposures (e.g., one meal or during one day). In these cases, the JMPR may establish an acute reference dose (acute RfD) that takes into account possible acute adverse effects. Because it is highly unlikely that an individual would consume two different commodities in large amounts within a short period of time and that both of those commodities would contain residues of the pesticide of interest and that those residues would be present at their respective maximum levels (i.e., MRLs), dietary exposure for assessing the risk of an acute hazard can generally be based on consumption of a large portion of a single commodity and on residues assumed to be at the MRL or at some other maximum level. In this regard, residue levels exceeding the MRL have been reported to occur on certain individual commodity units (e.g., one carrot) although all composite samples may be well below the MRL (MAFF, 1995). Consequently, data on residue levels in individual units may be required for certain commodities when a pesticide is considered to pose an acute hazard. If the short-term exposure assessment based on the large portion consumption of a food commodity does not exceed the acute RfD, it is highly unlikely that the acute RfD would be exceeded in practice; however, this procedure may need to be modified to accurately accommodate the toxicological findings. This approach also requires detailed data on food consumption by individuals, which can be obtained only at the national level. It has been proposed that GEMS/Food, in performing short-term dietary intake assessments, should use the highest reported daily intake from countries for each of the five GEMS/Food regions at the 97.5 percentile for only consumers of the food. Such large portion consumption data will be collected for the general population and for children age 6 and lower. Similarly, the 97.5 percentile residue level estimated from a distribution of residues on individual commodity units has been proposed for use in an acute hazard exposure assessment (FAO/WHO, 1997c).

GOOD AGRICULTURAL PRACTICE

GAP has a critical role in the formulation of draft MRLs proposed by the JMPR and elaborated by the CCPR. Guidelines developed by the CCPR (FAO/WHO, 1996b) encourage the use of more effective, but less persistent pesticides in order to reduce amounts of residues that can occur on food crops and in the environment. The guidelines recognize that differences in GAP exist among countries in the choice of pesticides, their use pattern, and residue levels on food. In order to deal with these differences, GAP for all countries from which data were submitted to the JMPR are used in the evaluation of supervised field trial residue data in developing draft MRL proposals. In effect, the JMPR's development of proposed draft MRLs and the CAC's elaboration of Codex MRLs lead to securing international agreement on GAP under widely varying conditions of pesticide use.

ELABORATION OF CODEX MRLs

An overview of the Codex step procedure for the elaboration of Codex MRLs is presented in Figure 13–2. Steps 1 and 2 have been discussed above. At step 3, draft MRLs proposed by the JMPR are introduced into the CCPR system for comment by member countries and international nongovernmental organizations (INGOs). The annual sessions of the CCPR serve as the forum for member countries to discuss each proposed draft MRL. Any country that objects to a proposed draft MRL is expected to give the basis for the objection, supported by scientific data. In such cases, the CCPR may refer draft MRL proposals back to the JMPR for further evaluation. An updated listing of the current status of draft MRLs in the Codex step procedure, including all existing Codex MRLs, is published annually by the Joint FAO/WHO Food Standards Programme in *Residues of Pesticides in Foods and Animal Feeds*.

At the end of the step procedure (i.e., step 8), draft MRLs are presented for adoption to the CAC, which meets every two years. After adoption, Codex MRLs are submitted to member countries for consideration in accordance with Codex acceptance procedures (FAO/WHO, 1997a). Guidance has been provided to national regulatory bodies to facilitate acceptance and use of Codex MRLs (FAO/WHO, 1986). Codex MRLs adopted by the CAC at step 8 are periodically published in English, French, and Spanish in the *Codex Alimentarius* (FAO/WHO, 1996).

Periodic Review

As part of its elaboration of Codex MRLs, the CCPR will place on the priority list for periodic review pesticides for which MRLs were first proposed over 10 years ago or for which a periodic review was conducted more than 10 years ago. Under this procedure, older pesticides are reevaluated by the JMPR to ensure that their ADIs and Codex MRLs are supported by current toxicological and GAP data. Since it was initiated in 1993, the periodic review procedure has resulted in Codex MRLs being

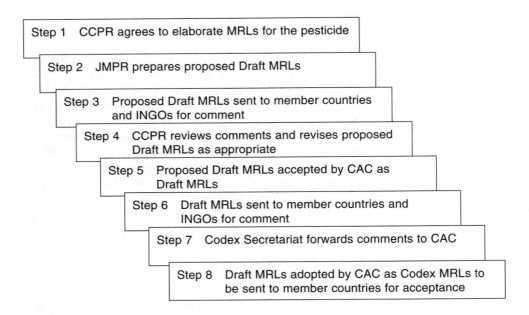

Figure 13–2 Codex Step Procedure for the Elaboration of Codex MRLs. CCPR indicates Codex Committee on Pesticide Residues; CAC, FAO/WHO Codex Alimentarius Commission; JMPR, Joint FAO/WHO Meeting on Pesticide Residues; and MRL, maximum residue limit. Courtesy of the World Health Organization, Geneva, Switzerland.

withdrawn for a number of pesticides. These withdrawals were due to lack of commitment to generate data to substantiate the continued use of the MRLs from either food safety or GAP perspectives. In other cases, ADIs and MRLs have been amended by the JMPR reevaluation. The amended MRLs are introduced at step 3 of the Codex step procedure. Pending the adoption of the amended MRLs at step 8, the original MRLs are maintained.

CODEX EXTRANEOUS RESIDUE LIMITS

Codex MRLs refer to pesticide residues arising from the purposeful and authorized use of a pesticide according to GAP. Pesticide residues also may arise in food commodities from environmental sources of contamination. In situations where these pesticide residues could affect international trade, the CCPR may elaborate an extraneous residue limit (ERL). Currently, Codex ERLs exist for certain older chlorinated pesticides that have had their registered agricultural uses withdrawn in most countries, but for which residues continue to persist in the environment. These include the pesticides aldrin and dieldrin, chlordane, DDT, endrin, and heptachlor.

Unlike an MRL, an ERL does not sanction or condone the use of a pesticide in food production. Instead, an ERL is based on the recognition that, even though their registered uses have been discontinued, residues of these pesticides on food cannot always be prevented. Draft ERL proposals developed by the JMPR are based on monitoring data submitted by GEMS/Food and national governments. Proposed ERLs are set no higher than necessary and should not accommodate purposeful use of the pesticide. For risk assessment purposes, JMPR may establish provisional tolerable daily intakes for these pesticides rather than ADIs. Elaboration of Codex ERLs follows the same step procedure as Codex MRLs (Figure 13-2).

INTERNATIONAL REGULATORY PRACTICES INVOLVING PESTICIDE RESIDUES

The work of the CCPR is not just restricted to the elaboration of MRLs and ERLs for pesticides. In fact, the CCPR over the years has developed a series of principles that may be viewed as international regulatory practices. These practices, which cover various regulatory issues, are important in ensuring uniformity and consistency in the application of Codex MRLs to food commodities in international trade. These international regulatory practices include a food classification system, MRL definitions of residue, portions of food to which Codex MRL apply, methods of food sampling, methods of analysis, good laboratory practice, and MRLs for processed food (FAO/WHO, 1993).

Problems Affecting the Use of Codex MRLs

The CAC through its CCPR offers the framework for countries to participate in and contribute to the formulation of internationally acceptable MRLs. Through the Codex step procedure and sessions of the CCPR, national governments have an opportunity to ensure that their safety concerns and agricultural needs are taken into account. There are, however, various factors that can impede governments from achieving this goal. The biennial reports on government acceptances of Codex MRLs have consistently indicated a very low rate of formal acceptances. Only a few countries, such as New Zealand and Singapore, have routinely adopted Codex MRLs into their national legislation. Even the United States, a strong supporter of the work of Codex, has accepted only about 35% of the Codex MRLs (GAO, 1991). The reasons for the low rate of formal acceptance of Codex MRLs vary. When considering the acceptance and use of a Codex MRL, member countries may encounter the following situations.

1. Codex MRL is lower than the national MRL. In this situation, an imported food that complies with the Codex MRL should also comply with the higher national MRLs. However, problems can occur for countries having national MRLs higher than those of Codex because their exports could contain residues in excess of the Codex MRLs and be prevented from entering foreign countries that use the lower Codex MRLs. These differences in Codex and national MRLs may result when the JMPR does not have access to residue data reflecting the GAP that requires the higher MRLs.

2. Codex MRL is higher than the national MRL. Member countries may perceive that accepting a Codex MRL higher than their national MRL would pose additional risk to their consumers although the actual increase in exposure may still be well below the level of no appreciable risk (i.e., the ADI). It may also be viewed as providing foreign producers with a production advantage in that they would be allowed to use more pesticide than domestic producers. The most common reason for a Codex MRL being higher than a national MRL is that some countries legitimately require a higher application rate and/or a shorter preharvest interval for effective pest control; these practices result in a higher MRL.
3. Codex MRL has no corresponding national MRL. Member countries may again perceive that accepting Codex MRLs for a pesticide use not registered in their country poses additional risks to consumers or is discriminatory to domestic food producers. A lack of a national MRL for a pesticide-food commodity combination with a Codex MRL may occur because the pesticide is not registered for the specific food use or because the pesticide is not registered for any agricultural uses. The former situation is easier to resolve as the national registration authority would have had reviewed the basic safety, agricultural use, and other data for the pesticide and found them acceptable.

Perhaps the greatest obstacle to greater government acceptance of Codex MRLs has been the lack of specific language in national legislation to allow their application to imported food. In the past, Codex MRLs have been viewed as advisory with no obligation on the part of member countries regarding their use. With ever-increasing international trade in food, however, concern over the potential use of health and safety requirements as nontariff barriers to trade has led to a landmark international agreement that may dramatically alter the way Codex MRLs are viewed in the future.

Codex MRLs and International Trade

With the establishment of the World Trade Organization (WTO) in 1995 and the coming into force of its Agreement on the Application of Sanitary and Phytosanitary Measures (SPS Agreement), Codex standards, guidelines, and recommendations, including Codex MRLs, have been recognized by WTO as representing the international consensus on requirements to protect human health from food-borne hazards (WTO, 1995). More importantly, for the first time, an international trade agreement explicitly recognized that establishment of rational harmonized regulations and standards for food in international trade must be based on a rigorous scientific process (i.e., risk assessment). The SPS Agreement also requires adherence to certain disciplines, including transparency, consistency, and nondiscriminatory application. By using Codex standards, guidelines, and recommendations, member countries will be considered as having complied with the requirements of the SPS Agreement. While the SPS Agreement does allow for countries to adopt other more strict requirements than those of Codex, countries will be required in trade disputes to provide the scientific justification for such deviations. Consequently, harmonization of national MRLs with those of Codex for foods in international trade is being given more emphasis by countries, and the deliberations of the CAC and the CCPR are considered of much greater importance.

Moreover, WHO and FAO have begun a critical examination of the application of risk analysis by the CAC and its subsidiary bodies as well as procedures used by its expert advisory bodies, such as JMPR in conducting risk assessments. Three consultations on risk analysis have been convened by WHO and FAO that have led to greater scientific rigor and consistency in the work of Codex (FAO/WHO, 1995; 1997d; and 1998). Coupled with the trade agreements of the WTO, these improvements in the work of Codex should lead to an increased acceptance of recommendations of Codex by member countries. In the future, greater recognition of Codex MRLs by member countries will contribute to the objective of CAC of protecting public health and facilitating international trade in food.

REFERENCES

Food and Agriculture Organization. (1997). *Manual on the Submission and Evaluation of Pesticide Residues Data for the Estimation of Maximum Residue Levels in Food and Feed.* Rome: Food and Agriculture Organization.

Food and Agriculture Organization/World Health Organization. (1982). *Results of Questionnaire on National Regulatory Systems for Pesticide Residues in Food, Codex Committee on Pesticide Residues.* CX/PR 81/15. Rome: Food and Agriculture Organization.

Food and Agriculture Organization/World Health Organization. (1986). *Guide to Codex Recommendations Concerning Pesticide Residues, Codex Committee on Pesticide Residues.* CAC/PR 9–1985. Rome: Food and Agriculture Organization.

Food and Agriculture Organization/World Health Organization. (1993). Pesticide residues in food. In *Codex Alimentarius*, Vol. 2, 2nd ed. Rome: Food and Agriculture Organization.

Food and Agriculture Organization/World Health Organization. (1995). *Application of Risk Analysis to Food Standards Issues.* Report of a Joint FAO/WHO Expert Consultation. Geneva, Switzerland, March 13–17, 1995. WHO/FNU/FOS/95.3. Geneva, Switzerland: World Health Organization.

Food and Agriculture Organization/World Health Organization. (1996). Pesticide residues in food—maximum residue limits. *Codex Alimentarius*, Volume 2B. Rome: Food and Agriculture Organization.

Food and Agriculture Organization/World Health Organization. (1997a). *Codex Alimentarius Commission Procedural Manual,* 10th ed. Rome: Food and Agriculture Organization.

Food and Agriculture Organization/World Health Organization. (1997b). Report of the 29th Session of the Codex Committee on Pesticide Residues. ALINORM 97/24A, para. 40. Rome: Food and Agriculture Organization.

Food and Agriculture Organization/World Health Organization. (1997c). *Food Consumption and Exposure Assessment to Chemicals.* Report of a Joint FAO/WHO Consultation, Geneva, Switzerland. February 10–14, 1997. WHO/FSF/FOS/97.5. Geneva, Switzerland: World Health Organization.

Food and Agriculture Organization/World Health Organization. (1997d). *Risk Management and Food Safety.* Report of a Joint FAO/WHO Expert Consultation. Rome, Italy, January 27–31, 1997, FAO Food and Nutrition Paper 65. Rome: Food and Agriculture Organization.

Food and Agriculture Organization/World Health Organization. (1998). *Risk Communication in Food Safety*. Report of a Joint FAO/WHO Expert Consultation. Rome, Italy, February 2–6, 1998. FAO Food and Nutrition Paper 70. Rome: Food and Agriculture Organization.

Government Accounting Office. (1991). *International Food Safety, Comparison of U.S. and Codex Pesticide Standards, Report to Congressional Requestors*. GAO/PEMD—91-2. Washington, DC: U.S. General Accounting Office.

Ministry of Agriculture, Fisheries and Food. (1995). *Consumer Risk Assessment of Insecticide Residues on Carrots*. York, United Kingdom: Pesticide Safety Directorate, Ministry of Agriculture, Fisheries and Food.

World Health Organization. (1953). *Toxic Hazards of Certain Pesticides*. WHO Monograph Series No. 16. Geneva, Switzerland: World Health Organization.

World Health Organization. (1982). *Constitution of the World Health Organization, Basic Documents,* 32nd ed. Geneva, Switzerland: World Health Organization.

World Health Organization. (1990). *Principles for the Toxicological Assessment of Pesticide Residues in Food*. Environmental Health Criteria 104. Geneva, Switzerland: World Health Organization.

World Health Organization. (1996). *Joint Meeting on Pesticides: Organisation and Structure, International Programme on Chemical Safety*. Geneva, Switzerland: World Health Organization.

World Health Organization. (1997). *Guidelines for Predicting Dietary Intake of Pesticide Residues, Global Environmental Monitoring System*. Document WHO/FSF/FOS/97.7. Geneva, Switzerland: World Health Organization.

World Trade Organization. (1995). *Agreement on the Application of Sanitary and Phytosanitary Measures, Legal Texts*. Geneva, Switzerland: World Trade Organization.

CHAPTER 14

Codex Standards for Veterinary Drug Residues

Palarp Sinhaseni and Richard J. Dawson

INTRODUCTION

In 1984, a Joint Food and Agriculture Organization/World Health Organization (FAO/WHO) Expert Consultation on Residues of Veterinary Drugs in Foods was held at the request of the 15th Session of the Codex Alimentarius Commission (CAC). This consultation recommended the formation of a new Codex Standing Committee on Veterinary Drug Residues in Foods (CCRVDF). This committee's formation was approved by the CAC in July 1985. At its first session, in 1986, definitions were agreed as follows:

- *Veterinary drug*—Any substances applied or administered to any food-producing animal, such as meat- or milk-producing animals, poultry, fish, or bees, whether used for therapeutic, prophylactic, or diagnostic purposes or for modification of physiological functions or behavior.
- *Residues of veterinary drugs*—Includes the parent compounds and/or their metabolites in any edible portion of the animal product, and includes residues of associated impurities of the veterinary drug concerned.

The terms of reference for the CCRVDF, approved by the CAC in 1985, are to

- Determine priorities for the consideration of residues of veterinary drugs in foods.
- Recommend maximum residue levels of such substances.
- Develop codes of practice as may be required.
- Determine criteria for analytical methods used for the control of veterinary drug residues in foods.

GENERAL PROCEDURES

Dr. L. M. Crawford (CAC, 1989) in the third session of the CCRVDF stated that "CCRVDF has developed international food standards that offer a high measure of protection and are practical and enforceable." The practicality and enforceability of the Codex standards are possible because the process of elaboration provides opportunities for comment from all interested parties, including government organizations, industry, and consumer representatives. Government employees as members of CCRVDF delegations represent their respective government organizations and policies, and serve as delegates and advisors. Members of a Codex committee delegation typically prepare government policy positions on agenda items. Trade, industry, and consumer group representatives participate as nongovernment advisors. Reports of the CCRVDF are a summary of deliberations and decisions made by the committee. A distinctive and common feature of Codex committee reports is that they must be unanimously accepted by all members before adjournment of the meetings. The review process leading to establishing a Codex standard can be triggered by direct requests to the FAO/WHO Secretariat from member countries as well as by requests from the CAC.

ELABORATION OF CCRVDF STANDARDS AND MAXIMUM RESIDUE LIMITS

The CCRVDF is responsible for recommending maximum residue limits (MRLs) for residues of veterinary drugs in food. It was agreed at the first session of the CCRVDF that for a candidate veterinary drug to be placed on the committee's priority list for the development of an acceptable residue level, it should meet some, but not necessarily all, of the following criteria:

1. The drug results in residues in the food commodity.
2. The drug or its residues are a matter of public health concern.
3. The residues of the drug affect international trade to a significant degree.
4. The residues of the drug are creating or having a potential to create commercial problems.
5. The drug is available for use as a commercial product.
6. It is used in accordance with good veterinary practice.

In addition, there must be a firm indication that relevant data will be made available for evaluation.

The Eighth Session of the Codex Committee on Residues of Veterinary Drugs in Foods (CAC, 1995) amended the criteria for the inclusion in, or exclusion from, the priority list as follows:

1. Use of the drug has potential to cause public health and/or trade problems.
2. The drug is available as a commercial product.
3. A dossier will be made available.

Adoption as a Codex Standard on Veterinary Drug Residues can be attained after completion of the following eight-step process:

- Steps 1–3—The Codex Secretariat refers compounds identified by the CCRVDF for evaluation to the Joint FAO/WHO Expert Committee on Food Additives (JECFA), an independent body that conducts reviews on substances for governments, industries, and consumers, as well as Codex. JECFA recommendations for MRLs for veterinary drug residues are then distributed for comments from governments and interested international organizations on all aspects, including possible implications of the draft recommendations for maximum limits for veterinary drug residues on their economic interests.
- Step 4—The CCRVDF examines the recommendations for maximum limits for veterinary drug residues in the light of received comments. The Codex committee indicates to the CAC proposed draft maximum limits that need to be passed through the full procedure and those for which steps 6 and 7 might be omitted when completion of the standard is a matter of exceptional urgency, or when the standard is uncontroversial and has already proved to be generally acceptable to members of the committee.
- Steps 5–8—The draft standard is submitted to the CAC, with any written proposals received from members for amendments at step 8, with a view to its adoption as a Codex standard on veterinary drug residues.

JOINT FAO/WHO EXPERT COMMITTEE ON FOOD ADDITIVES

After the Consultation on Residues of Veterinary Drugs in Foods in 1984 (see "Introduction"), the CAC called upon FAO and WHO to consider the formation of an appropriate expert body to provide independent scientific advice to the committee.

Relationship with the JECFA

In its first session, the CCRVDF confirmed that it wished to receive from the expert committee recommendations for acceptable residue levels of individual veterinary drugs in specific foods. The committee felt that it was appropriate to work with an expert committee for the evaluation of veterinary drugs in foods, and a relationship with JECFA was formed. Temporary experts who work in the various JECFA sessions are selected from the scientific community on the basis of their scientific expertise on the subject matter being reviewed. They work as individuals and not as representatives of their parent organizations, be they public or private associations. Consultants and advisers are engaged to review the literature for the committee and provide members with synopses. To date, JECFA has reviewed veterinary drugs at 10 of its sessions.

Figure 14–1 summarizes the relationships between JECFA, CCRVDF, and national governments.

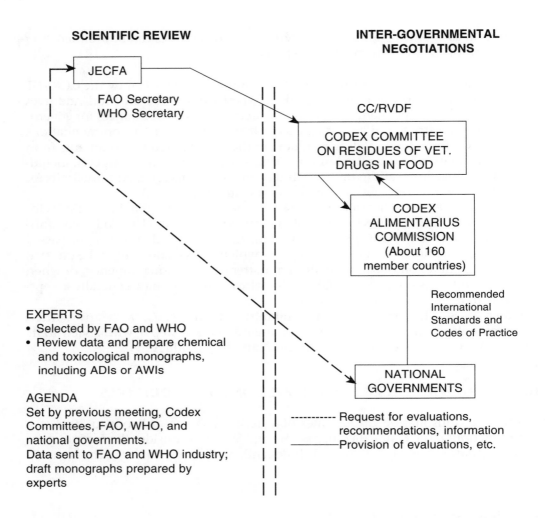

Figure 14–1 Relationships between JECFA, CCRVDF, and National Governments during the Procedure for the Elaboration of Codex MRLs for Veterinary Drugs. *Source:* Copyright © 1997, Raj Malik.

During CCRVDF's 2nd meeting in 1987, the committee confirmed the concept that establishing Codex recommendations for residue levels of veterinary drugs in foods should be based primarily on a safety evaluation of toxicological data, and should take into account good veterinary practice in the use of veterinary drugs as well as factors such as available methods of analysis.

Relevant Data for Assessing the Human Food Safety of Residues of Veterinary Drugs

In recommending an MRL for a specific compound, several factors are taken into account by the JECFA. Among them are the results of toxicological and radiolabel residue studies, the bioavailability of bound residues, the identification of target tissue(s), the existence of a residue marker to determine compliance with safe residue limits, residue data from use of the veterinary drug according to good practice in the use of veterinary drugs, withdrawal periods for adequate residue depletion of the drug, and practical methods for residue analysis.

When evaluating the toxicological aspects of residues of veterinary drugs in food, the JECFA requires detailed reports (including individual animal data) from the following types of studies:

- pharmacokinetic, metabolic, and pharmacodynamic studies in experimental and food-producing animals, and in humans when available
- Short-term, long-term/carcinogenicity, reproduction, and developmental studies in experimental animals, and genotoxicity studies
- special studies designed to investigate specific effects, such as those on mechanisms of toxicity, no-hormonal–effect levels, immune responses, or macromolecular binding
- for compounds with antimicrobial activity, studies by the manufacturer designed to evaluate the possibility that the compound might have an adverse effect on the microbial ecology of the human intestinal tract
- studies on the use of, and exposure to, the drug in humans, including studies of effects observed after occupational exposure and epidemiological data following clinical use in humans

Detailed reports of studies relevant to the evaluation of drug residues in food-producing animals that are required for evaluation include information on:

- the chemical identity and properties of the drug
- its use and dose range
- as for the toxicological evaluation, pharmacokinetic and metabolic studies, in experimental animals, target animals, and humans when available

The principles governing the safety evaluation of residues of veterinary drugs in food by JECFA have been elucidated (Cerniglia, 1995 and WHO, 1987; 1988; 1989; and 1990), in particular in relation to bound residues, microbiological risks, and allergenicity.

JECFA members carry out peer reviews of the published literature and any available unpublished industry data. The reviewed material is discussed at JECFA meetings, and recommendations are developed for CCRVDF. Typically, when performing a review on a substance, residue depletion data are invaluable when recommending MRLs. The availability and performance of analytical methods are thoroughly reviewed, which may play a critical role in experts' recommendations for MRLs. Another important factor to be considered in establishing MRLs in edible animal products is the amount of the food item consumed.

Procedures for utilizing food consumption factors and daily intake data to determine theoretical maximum residue exposures were considered in the 34th, 40th, 42nd, and 45th meetings of JECFA (WHO, 1989; 1993; and 1995). The committees use intake data at the upper limit of the consumption range for individual animal products. Its recommendations on the daily intake values to be used were: 300 g of meat (as muscle tissue), 100 g of liver, 50 g of kidney, 50 g of tissue fat, 100 g of egg, and 1.5 l of milk. JECFA encouraged governments to consider whether local diets might result in intakes that exceed the allowable daily intake value. However, JECFA considered this to be a very rare occurrence (WHO, 1993).

SELECTION OF ANALYTICAL METHODS AND SAMPLING FOR THE CONTROL OF VETERINARY DRUG RESIDUES IN FOODS

One of the mandates of CCRVDF, in its terms of reference, is to determine analytical methods used for the control of veterinary drug residues in foods. Practical and enforceable food standards cannot be achieved without work of CCRVDF.

The needs of governments for analytical methods with good reliability, and the highly specialized expertise needed in dealing with methods of analysis and sampling for residues of veterinary drugs were accepted by CCRVDF in its first meeting in 1987. At that meeting, the CCRVDF established an ad hoc Working Group on Methods of Analysis and Sampling to elaborate and recommend appropriate methods of analysis and sampling to the plenary session of CCRVDF. This working group's role must be reaffirmed at each CCRVDF meeting.

The CCRVDF recommends methods that have been demonstrated to be suitable for use in regulatory laboratories, where analyses are conducted in support of MRLs that meet the Codex criteria. Analytical methods suitable for use in such regulatory work (MacNeil, 1998) should take account of the following:

- The method is practical to use in a routine residue laboratory, requiring no unusual (very expensive or not commercially available) equipment and having an elapsed time from start to finish of an analytical run of less than two working days.
- The reagents required should be commercially available.
- Critical control points in the method should be identified, and the method's applicability should be clearly stated.
- The method should not use solvents banned under the Montreal Protocol.
- The method should include clear worker safety instructions.
- A written protocol giving all information required to apply the method in a laboratory should be available. This should include a statement of expected performance standards.

CONCLUSION

International food standards are desirable when they offer a high measure of protection and are practical and enforceable for all countries. The Codex Standards on Veterinary Drug Residues can be achieved through the elaboration of Codex MRLs for veterinary drugs. Through complementary interaction among JECFA,

CCRVDF, and national governments, the desired Codex standards can be attained. The CAC, at its 20th session, agreed that the interactive framework should be developed further. Coordinated effort was encouraged to harmonize risk assessment in relation to the continued establishment of FAO/WHO Codex Standards on Veterinary Drug Residues.

The assessment procedures used by JECFA are continually evolving. The adopted procedures reflect the current dynamic nature of scientific knowledge and greater experience of the JECFA committee. To ensure an adequate supply of safe, wholesome, and nutritious food, these standards have to be considered in relation to whether they can be accepted and can be applied in the member countries within specific local, social, and economical contexts.

REFERENCES

Cerniglia, C. E. (1995). *Assessing the Effects of Antimicrobial Residues in Food on the Human Intestinal Microflora*. JECFA CI 1995/1. Rome: Food and Agriculture Organization.

Codex Alimentarius Commission. (1989). *Joint FAO/WHO Food Standards Programme, Eighteenth Session, Geneva, 3–14 July 1990*. Report of the Third Session of the Codex Committee on Residues of Veterinary Drugs in Foods. Washington, DC, October 31–November 4, 1988. ALINORM 89/31A. Rome: Codex Alimentarius Commission.

Codex Alimentarius Commission. (1995). *Report of the Eighth Session of the Codex Committee on Residues of Veterinary Drugs in Foods*. Washington, DC, June 7–10, 1994. ALINORM 89/31. Rome: Codex Alimentarius Commission.

Malik, R. (1997). *Food Safety and Risk Analysis*. A training course organized for trainees from the Asia and Pacific countries by FAO and ILSI. Chiang Rai, Thailand, December 14, 1997.

MacNeil, J. D. (1998). *JECFA Requirements for Validation of Analytical Methods*. Working Paper Prepared for Consideration by the 50th Joint FAO/WHO Expert Committee on Food Additives, Rome, February 1998.

World Health Organization. (1987). *Principles for the Assessment of Food Additives and Contaminants in Food*. WHO Environmental Health Criteria, No. 70. Geneva, Switzerland: World Health Organization.

World Health Organization. (1988). *Evaluation of Certain Veterinary Drug Residues in Food*. Thirty-Second Report of the Joint FAO/WHO Expert Committee on Food Additives. WHO Technical Report Series No. 763. Geneva, Switzerland: World Health Organization.

World Health Organization. (1989). *Evaluation of Certain Veterinary Drug Residues in Food*. Thirty-Fourth Report of the Joint FAO/WHO Expert Committee on Food Additives. WHO Technical Report Series No. 788. Geneva, Switzerland: World Health Organization.

World Health Organization. (1990). *Evaluation of Certain Veterinary Drug Residues in Food*. Thirty-Sixth Report of the Joint FAO/WHO Expert Committee on Food Additives. WHO Technical Report Series No. 799. Geneva, Switzerland: World Health Organization.

World Health Organization. (1993). *Evaluation of Certain Veterinary Drug Residues in Food*. Fortieth Report of the Joint FAO/WHO Expert Committee on Food Additives. WHO Technical Report Series No. 832. Geneva, Switzerland: World Health Organization.

World Health Organization. (1995). *Evaluation of Certain Veterinary Drug Residues in Food*. Forty-Second Report of the Joint FAO/WHO Expert Committee on Food Additives. WHO Technical Report Series No. 851. Geneva, Switzerland: World Health Organization.

CHAPTER 15

Development of Radiological Standards

Stuart Conney and David Webbe-Wood

INTRODUCTION

The practice of radiation protection has been around almost as long as the use and study of the phenomenon of radiation. In the context of this book, the term *radiation* means ionizing radiation; nonionizing radiation (e.g., ultraviolet) is also important to human health, but not in the contamination of foodstuffs. In the early days, when the use of radiation principally involved the external exposure of, for example, a patient to diagnostic radiographs, the main concerns about health were about the acute effect of high doses to medical practitioners. The long-term effects of cancer inducement were not recognized until later when the effects of these early high-dose exposures were seen in increased mortality.

There are different forms of radioactivity and radiation, but they all affect the body in similar ways. The damage they do can be expressed in the concept of radiation dose, a measure of harm done to the body's tissues. Radioactive contamination of food is just one of the ways a person can be exposed to radiation, and the effects to the consumer can also be quantified as radiation dose. Therefore, the risk assessment, regulation, and legislation of radioactivity in food have to fall under the more general framework of radiation protection (i.e., the control of radiation dose to people).

In this chapter, we will discuss the development of radiation protection as a whole and consider how the contamination of food and feeding stuffs fits into this framework. The various international, regional, and national organizations whose remit it is to develop standards in radiation protection, and hence standards for contaminants in foods, are then discussed, along with their interrelationships and some of the limitations this system produces. The methods used in assessing the effects of radioactivity on the food chain are considered along with the ultimate potential effects of the exposure to radiation on the consumer. The regulations governing the control of potentially contaminated foods are considered in relation to the bodies that implement such legislation and are put in the context of the other pressures involved in framing such laws.

In order to illustrate how standards are set, the United Kingdom is focused on in its regional and international context. Other countries, in different regions or trading conditions, will have a slightly different emphasis regarding how the standards are framed; but the basic concepts of public protection are set in the international context. Interpretative differences are needed to account for national and regional variations of how humans exploit the environment to produce food and in what quantities these foods are eaten.

RADIOACTIVITY AND RADIATION PROTECTION

Radioactive contaminants differ from chemical and biological contaminants of foods in a number of fundamental ways. For example, at the levels of radioactive contaminants seen in the environment from discharges from the nuclear industry, hospitals, and other institutions, there is no perceptible change or effect in the foodstuffs, or on the living system that produces it. The levels of plutonium-239, for example, seen in fish from near the Sellafield site, in the Irish Sea, are measured at about 3 parts per million billion (MAFF, 1996), while detection limits are lower still—far lower than those of other contaminants.

All elements are capable of having radioactive isotopes, which may be discharged into the environment as wastes from the various processes where radioactivity is used. These include power generation, including fuel reprocessing; manufacturing radioactive sources for medical uses; and processes not directly involving the use of radioactivity, but where the process coincidentally enhances levels of naturally occurring radioactivity. Each element has its own chemical and physical properties that determine how it interacts with the various food-producing processes in terrestrial and aquatic environments. The chemical form, in particular, is dependent upon the process that introduced the radioactivity into the environment and is not always easily determined at the low concentrations involved.

In addition, humans are exposed to substantially greater natural background radiation than to anthropogenic (man-made) sources (Figure 15–1). The natural background radiation is made up of cosmic rays, naturally radioactive substances in rocks, radon gas, and naturally occurring radioactive substances in the human body and diet (Hughes & O'Riordan, 1993).

Individuals may come into contact with radiation in a number of ways. For example, they may be exposed to a source of ionizing radiation, such as an X-ray machine, or radioactivity may also be taken into the body through ingestion or inhalation, or via a wound. Removal of the source of radiation is simple, and doses can be controlled once the risk has been recognized. Once radioactive elements have been absorbed by the body, the situation is more complex with the behaviors being dependent on the chemical characteristics of the material involved. Biological half-times vary from a few days to far longer than the normal human life span. Therefore, for protection purposes, all the energy deposition in the body is conservatively assumed to occur in the first year it was absorbed into the body.

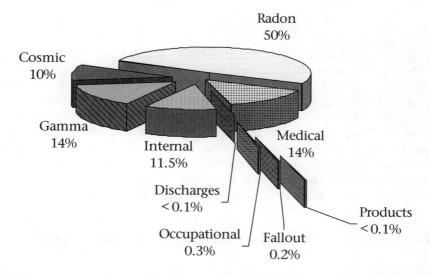

Figure 15–1 Average Annual Dose to the U.K. Population—2.6 mSv Overall. *Source:* Reprinted with permission from J.S. Hughes and M.C. O'Riordan, *Radiation Exposure of the UK Population—1993 Review*, NRPB-R263, 1993, National Radiological Protection Board.

Radiation may affect the body in a number of ways. Acute effects, caused by large doses over a short period, are generally the result of cell death, which occurs during cell division. It follows that tissues with rapidly dividing cell populations, such as bone marrow and skin, are more sensitive to this type of radiation damage. There is a well-defined threshold of exposure, many times natural background, below which these effects do not occur. Small doses may not kill cells, but may damage the genetic material within the cell, leading to an increased risk of cancer in the exposed individual or potential genetic mutations in future generations. These effects, by their nature, occur sometime after exposure, against a natural background in the population as a whole and have no threshold. The risk is proportional to the dose of radiation received. As one radiation impact may damage genetic material, a larger number increases the chances of mutation.

The basic system of radiation protection has developed into the principles used today. These involve three stages: (1) a process of justification, (2) optimization where doses are kept as low as reasonably achievable (ALARA), and (3) adherence to limits (ICRP, 1991). The first part of this process will usually include a measure of economic judgment as well as a decision on whether a process is required and whether there is a net benefit to society. Optimization, in this context, is where operations are implemented in such a way as to maximize the net benefit. The limits are set for the total doses from controlled sources of man-made radiation (excluding medical doses to patients). To enable exposures from different isotopes to be summed, dose coefficients for radionuclides are published (IAEA, 1994 and ICRP, 1995). These dose coefficients take into account the differing behavior in the body of different radionuclides together with the particular properties of the emitted radiation. Thus, doses from external radiation can be summed with doses from the intakes of radionuclides to give a measure of the overall risk to the health of the exposed individual. This is unlike most other contaminants where there is no method of judging an exposure against a mortality risk. This dose-response relationship has been built up from data on atomic bomb survivors and other groups exposed to known doses of radiation.

In the most recent recommendations of the International Commission on Radiological Protection (ICRP) (ICRP, 1991), a new concept of dose constraint was developed. This is a level of dose, less than the dose limit, below which exposure from new facilities should be kept. These are primarily a design tool to enable the optimization of doses from planned facilities at the design stage.

The foregoing is only a brief outline of some of the considerations in radiation protection. Several texts deal with the subject; a good primer on this subject is *Introduction to Radiation Protection* (Martin & Harbison, 1996).

INTERNATIONAL COMMISSION ON RADIOLOGICAL PROTECTION

The ICRP was established, under the name of the International X-ray and Radium Protection Commission, in 1928. In 1950 the ICRP was restructured, and the current name was adopted. The ICRP has no formal powers. It is a self-elected body of experts that makes recommendations. These recommendations are generally accepted by most governments and by international bodies such as the European Union (EU). It is these governments and bodies who are responsible for the production of regulations.

Over the years, the interests of the ICRP have developed to incorporate a wide range of uses of ionizing radiation and practices that result in the generation of radiation and radioactive material. Official relationships are maintained by the ICRP with the World Health Organization (WHO) and the International Atomic Energy Agency (IAEA). Relations are also maintained with a number of other organizations. Funding is provided by national and international organizations. The ICRP has established four committees, and members of these committees form task groups to investigate specific areas. Currently the committees and task groups are as follows:

- Committee 1, Radiation Effects—This committee considers the risks of induction of cancer and heritable diseases together with the underlying mechanisms of radiation action; and the risks, severity, and mechanisms of induction of tissue/organ damage and developmental defects.
- Committee 2, Doses from Radiation Exposure—This committee is concerned with developing dose coefficients for the assessment of internal and external radiation exposure; development of reference biokinetic and dosimetric models; and reference data for workers and members of the public.
- Committee 3, Protection in Medicine—Committee 3 is concerned with human protection (including unborn children) when ionizing radiation is used for medical diagnostics, therapy, or biomedical research.
- Committee 4, Applications of Recommendations—This committee is concerned with providing advice on the application of the recommended system in all its facets for occupational and public exposure. It also acts as the major point

of contact with other international organizations and professional societies with protection against ionizing radiation.

These task groups function by considering available information and use their expertise to apply that information to the circumstances under consideration. In many cases, they are limited by the availability of data. Much of the work on risk factors is based on data from atomic bomb survivors or from medical exposure. In both cases, these are high-dose, high-dose rate exposures, whereas exposure under consideration by ICRP is generally low-dose, low-dose rate. Therefore, both extrapolation and the expertise of ICRP members are required to derive risk estimates. Similarly, the evaluation of dose-per-unit-intake coefficients—to relate the dose received by an individual to the amount of activity ingested or inhaled—requires extrapolation from animal and human experiments and carries the proviso that if better data are available, then these should be used (e.g., the derivation of site-specific gut transfer factors for americium in winkles in the vicinity of Sellafield [Hunt et al., 1990]). The recommendations of the ICRP are published in a series of publications that commenced in 1959 and numbered 74 by 1998.

Over the decades, there has been a change in the emphasis in presentation of the ICRP's recommendations. In the early history of the ICRP, the tendency was to regard compliance, with dose limits, as representing a satisfactory level of achievement. Over the years, the ICRP's recommendations for the limit for whole-body exposure have been revised downward several times (Table 15–1). Also the ICRP's estimates of risk factors have changed over time, most recently with the latest recommendations (ICRP, 1991). In more recent times, the emphasis has moved to a position where the limits represent an upper bound that should not be exceeded. The emphasis is being placed on restricting doses to an ALARA level with economic and social factors being taken into account. (Previously, the emphasis was on numerical limits.) This change in emphasis has resulted in lower doses.

As the recommendations of the ICRP are applied in a wide range of circumstances and by a range of authorities and organizations, the degree of detail within the recommendations is deliberately restricted. The ICRP has no power to produce regulations. It is the responsibility of governments (including bodies such as the EU) to produce regulations to protect the environment of their national territories and their populations. Most governments follow the recommendations of the ICRP and use them to develop their own legislation, regulation, and licenses.

Table 15–1 Development by the ICRP of Dose Limits for the General Public

Year	Population Exposed	Whole Body Exposure Limit	Source
1954	Public	1.5 rem (15 mSv), 1/10 occupational dose	Caufield, 1989
1958	Public in vicinity of controlled area	0.5 rem (5 mSv)	ICRP, 1958 and ICRP, 1959
1977	Public	5 mSv	ICRP, 1977
1985	Public	1 mSv	ICRP, 1985
1990	Public	1 mSv (averaged over 5 years)	ICRP, 1991

The ICRP has made no recommendations on the levels of radionuclides in foodstuffs with the exception of recommendations made for intervention following an accident (ICRP, 1993). The ICRP approach to the protection of the general public (i.e., those who are not receiving the dose as a consequence of their employment or from medical procedures) from controlled sources of radiation of man-made origin is based on the concept of the critical group. *Critical group* is defined as those representatives of the population who because of their location or habits are most exposed to radiation as a consequence of the practice under consideration. Doses from all pathways are evaluated for this group (e.g., ingestion of foods containing radionuclides, direct irradiation from the site, or inhalation of radioactive gases and particles), and the total dose is evaluated. In this way, potentially complex and wide-ranging systems of exposure to radiation and radioactive materials can be combined to give one overall risk. The ICRP recommends that a limit of 1 mSv (milliSievert) per year should apply with a higher level being permitted in special circumstances, provided that the average over a five-year period does not exceed 1 mSv per year. This corresponds to a risk of developing a fatal cancer of 1 in 20,000 and of hereditary effects of 1 in 75,000. The task of refining these recommendations into a practical system of control, which may be incorporated into legislation, falls on other international and national bodies as described below.

UNITED NATIONS SCIENTIFIC COMMITTEE ON THE EFFECTS OF ATOMIC RADIATION

The United Nations Scientific Committee on the Effects of Atomic Radiation (UNSCEAR) was established in 1955 by the 10th General Assembly of the United Nations. The aim of the ICRP is to assess the consequences to human health of ionizing radiation and to estimate the dose to people all over the world from natural and man-made radiation sources. Twenty-one member countries of the United Nations are represented on the ICRP together with representatives of the WHO, the United Nations Environment Programme, the IAEA, the International Commission on Radiological Protection, and the International Commission on Radiation Units and Measurements.

UNSCEAR makes no recommendations on acceptable levels of radionuclides in foods. Its interests in foods are limited to the evaluation of doses from all sources and all pathways including ingestion via the food chain.

INTERNATIONAL ATOMIC ENERGY AGENCY

The IAEA was set up in July 1957 following a United Nations Conference the previous year. Later in 1957, the UN General Assembly approved its relationship with IAEA. Under its statute, the IAEA has two principal aims. The first is to "seek to accelerate and enlarge the contribution of atomic energy to peace, health and prosperity throughout the world." It is also tasked with ensuring that any assistance given by, or channeled through it, is not used in such a way as to further military purposes (Union of International Associations, 1994).

The constitution of the IAEA includes member states that signed and ratified the 1957 statute within 90 days or those that have since provided an instrument of acceptance after their membership was approved by the General Conference on the recommendation of the Board of Governors. The Board of Governors meets quarterly and the General Conference meets annually at the Vienna headquarters.

The day-to-day running of the IAEA is undertaken by a Secretariat of some 2,000, headed by the director general. This is split into six divisions covering administration, nuclear energy, research and isotopes, safeguards, nuclear safety, and technical cooperation—each headed by a deputy director general. Responsibility for safety standards rests with the nuclear safety division.

In addition to its headquarters in Vienna, IAEA also runs a number of laboratories that play a greater or lesser part in the setting of radiation standards and associated quantities. The laboratories at Siebersdorf include an agricultural laboratory, operated jointly with the Food and Agriculture Organization (FAO), and the physics and chemistry laboratory, which analyzes samples of foods for their radioactive content. The IAEA Marine Environment Laboratory at Monaco carries out studies on method development and marine processes to aid modeling of radioactivity through the marine food chain. The IAEA also employs expert groups to study and collate the most up-to-date information available on the processes involved in radioactive contamination of the food chain back to man. This information is made available to all states.

The IAEA first published *The Agency's Health and Safety Measures* (IAEA, 1960) in 1960 with a reissue in 1976. These documents stated that the safety standards would be based to the fullest extent possible on the ICRP recommendations. The principal methods of promulgating standards in nuclear safety for the public, workers, and those exposed to medical radiation are the basic safety standards (BSS) that have been produced in 1967, 1982 (new edition), and 1996 (IAEA, 1967 and 1996), following closely the relevant *Recommendations of the International Commission on Radiological Protection* (ICRP, 1959, 1977 and 1991). It should be noted that both the ICRP and its sister commission, the ICRU, have consultative status under agency rules. Other organizations cooperate regularly with the IAEA. These include the FAO and WHO, organizations of the United Nations.

In 1986 IAEA expanded its nuclear safety-related activities, and more recently the Nuclear Safety Division was separated from the Nuclear Energy Division.

The new BSS, based on 1990 recommendations of the ICRP (ICRP, 1991), were published by the IAEA in 1996 in the Safety Series No. 115, replacing those adopted in 1982 based on 1977 recommendations of the ICRP.

In line with the ICRP, the BSS mentions the use of constraints in optimizing exposure to radiation. Like the ICRP, no levels for constraints are suggested—this is left up to the individual state. Another area where the BSS provides guidance, but no hard and fast rules, is over intervention following a radiological emergency. The BSS states that set dose limits will not apply in such cases. The approach taken is that all interventions have a cost and that this must be balanced against the cost of the dose to exposed persons. This leaves post-Chernobyl interventions unaffected.

The IAEA is also involved in assessing doses received by the general public from consuming contaminated foods. As well as the BSS, published in the Safety Series, data necessary to calculate the doses from contaminated food are published in the technical report series. These include concentration factors, which enable the concentrations in aquatic biota to be calculated, and models that enable the transfer of radionuclides through the terrestrial environment to be followed. The IAEA also publishes generic, consumption rate data to enable member countries and others to assess annual intakes of radionuclides, given the concentration in the foods. These generic data are for use when no better local data are available. The data are reviewed by a group of experts from the member countries prior to publication by IAEA. The data provided on the dose-per-unit intake by the ICRP (ICRP, 1995) are recommended for use in dose calculations by the IAEA.

The IAEA are involved in a number of areas associated with the levels of radioactivity occurring in the food chain. For example, at the end of 1993, it initiated a program to collect and review environmental monitoring data (IAEA, 1995), in collaboration with UNSCEAR and WHO. This led to a database of radioactivity in food and other environmental materials under normal circumstances. Following the Chernobyl accident in 1986, research into the effects of the accident on the local food chain and environment was coordinated by the IAEA; IAEA acted as a conduit for advice and aid on the effects of the accident in the former Soviet Union.

Most governments tend to follow the recommendations of the IAEA but, due to local circumstances, the spirit of the recommendations is followed rather than the letter. The BSS and other safety standards derived by the agency are issued in the form of regulations and codes of practice that member states may take up and use in their own legislation.

EUROPEAN ATOMIC ENERGY COMMUNITY/EUROPEAN COMMISSION

The European Atomic Energy Community (EURATOM) came into existence with the signing of the EURATOM Treaty at the same time as the better known Treaty of Rome in 1957 (U.K. Government, 1972). These remained three separate communities until the Merger Treaty, which established a single council and a single commission, was signed in 1965. There have been other treaties since that have admitted further members, or changed budgetary arrangements, but the basic constitution has remained constant. The treaty, in common with several signed at this time, was designed to promote the fledgling nuclear industry in Europe, but it also contained articles promoting the health and safety of the peoples of the signatory states. The first of these health and safety articles stated that BSS should be laid down which, for the general public, would consider the maximum permissible doses and levels of exposure and contamination. Other articles set out how member states would implement the standards, monitor radioactivity levels, and supply information on a proposed new plant and its health and safety effects on neighboring states in the community.

Once the work of EURATOM was subsumed into the work of the European Community (now the EU), as a whole the organization of its work was absorbed by the European Commission and the Council of Ministers (the council). Administration of the European Commission's work is carried out by civil servants who are grouped under directorates-general (DGs), who work with one of the commissioners. Environment, Nuclear Safety, and Civil Protection comes under DG XI (see Chapter 4, European Community Legislation on Limits for Additives and Contaminants in Food).

European Commission staff will generally start to review the BSS because they are stimulated by a change in the advice from outside bodies (e.g., ICRP). The amendment of the BSS is carried out by a group of scientific and technical experts drawn from EU member states; some of these experts specialize in public health issues. The group, with a secretariat from the European Commission, produces its recommendations, which are then passed to the Social and Economic Committee for further comment—these stages are laid out in the treaty. The Social and Economic Committee then forwards the completed recommendations to the European Parliament for their comments and amendments. The completed document is then passed to the European Council of Ministers for adoption. Once BSS are adopted, they take the form of a directive, essentially the same as any other piece of EU legislation, which binds EU member states to the objectives, but leaves national governments to choose the form and means of use. A directive will generally stipulate time frames by which it should be included in individual states' legislation.

Since the EURATOM Treaty was signed in 1957, there have been various issues and amendments to the BSS. The first directive that mandated BSS was adopted in 1959. BSS were amended in 1962 and 1966 (EC, 1996) as the effects of ionizing radiation on the human body became known. Further amendments followed changes to dose-per-unit-uptake figures and a major review by the ICRP of its recommendations. As with most radiation safety recommendations, since all radiation effects are covered by the BSS, amendments to the BSS may have little effect on the levels needed to approach the currently recommended dose limit. Further amendments of the BSS were carried out in 1976, 1979, 1980, and 1984 (EC, 1976; 1979; 1980; and 1984).

Following publication of the recommendations of the ICRP (ICRP, 1991), the European Commission started the process of reviewing the BSS by consulting with technical and scientific experts, appointed by the Scientific and Technical Committee, from member states. The previous version of the BSS dated back to 1980, with an update in 1984, which were based on the previous recommendations in 1977 (ICRP, 1977). The first part of the consultation process was to obtain the opinion of the Economic and Social Committee. The draft BSS were passed to the Economic and Social Committee in July 1992, which adopted its opinion in February 1993. The next stage was to put the proposed BSS before the European Parliament. This was done with an amended version being produced in April 1994. The BSS were issued by the European Council as a council directive in May 1996 (EC, 1996).

There are a number of specific areas where the BSS take matters further than the ICRP recommendations, thereby reflecting other bodies of opinion. A notable area is the exemption from regulatory control of certain quantities or activity concentrations. These are specified for about 300 radionuclides where the chemical speciation is assumed to be the most radiotoxic form.

In line with the ICRP, the BSS mention the use of constraints in the optimization of exposure to radiation. It, like the recommendations of the ICRP, also suggests no level of constraint, but leaves this to the individual state. Another area where BSS provide guidance, but no hard and fast rules, is over intervention following a radiological emergency. The BSS state that the dose limits set out in the document will not apply in such cases. The BSS take the approach that all interventions have a cost and that this must be balanced against the cost of the dose to exposed persons. Such an approach, of course, leaves the post-Chernobyl interventions unaffected. The respective directive (EC, 1996) also allows member states to legislate stricter laws on radiation protection, but they should inform the European Commission and other member states. The deadline to implement this directive is May 2000, and for the first time all previous directives are repealed.

LIMITATIONS OF INTERNATIONAL BODIES

The international bodies referred to above are not without their limitations. Much of the effort of these organizations has been directed to the control of radiation exposure in the workplace and during medical procedures. Little direct consideration has been given to the limitation of radioactivity levels in foods.

Criticism has been leveled at the composition of the international bodies. In particular, it has been alleged that they are dominated by those employed by the nuclear industry or by organizations dependent on the nuclear industry for funding, with no representation from trade unions or consumer groups. Indeed, this criticism has been extended (Bertell, 1985) to relate to the profession of health physics as a whole. Criticism has also been leveled at the racial and gender composition of the ICRP and its task groups.

There has also been criticism (Bertell, 1985) that studies into the effect of radiation have been limited to considering cancer induction and that insufficient attention has been given to other impacts such as birth defects, depression of the immune system, and asthma. It has also been claimed that because the majority of the research and data relating to the health effects of ionizing radiation originates from organizations connected with the nuclear industry, it is impossible for independent scientists to challenge the conclusions of the ICRP.

It has been suggested (Edwards, 1997) that radiation may induce genomic instability, resulting in defects in future generations and that this effect has been given insufficient consideration by national and international bodies.

The role of the ICRP is to produce recommendations based on a thorough review of the available information and data. These reviews are carried out by scientists eminent within the field. Any lack of diversity in these scientists is a reflection of inequalities within society as a whole and does not represent an attempt to disenfranchise any group within society. The responsibility for producing legislation as a result of ICRP recommendations lies with governments. It is the responsibility of governments, not the ICRP, to consider, and if appropriate, reflect the concerns of trade unions and other stakeholders.

Criticism has been leveled at the ICRP that on occasions its recommendations are too restrictive and that it has not given due consideration to evidence that there may be a threshold in the dose/response relationship. There is a significant delay between recommendations being made and their being incorporated into national laws. For example, the 1990 recommendations of the ICRP will not be incorporated into United Kingdom legislation until 2000 at the earliest following the acceptance of the EU BSS, which specify many of the data values to be used in assessing doses.

ACCIDENT REGULATIONS

Consideration has been given to the control of foods contaminated as a result of an unplanned release from the early days of the nuclear industry. Prior to a fire at Windscale in the United Kingdom in 1957, the Ministry of Agriculture, Fisheries and Food (MAFF) (see "Enforcement in the United Kingdom") had considered the likely pathways back to man from the food chain. However, permissible levels in milk were calculated as the accident developed, finally resulting in a concentration of 3,700 becquerels (Bq)/l of iodine-131 in milk. These were not formally adopted in legislation, and developments in assessing radiation dose meant that several different levels were seen in the following years (Arnold, 1995).

Following the reactor accident at Chernobyl in April 1986, several countries were affected, to a greater or lesser extent, by fallout from the resultant cloud of radioisotopes. The initial response in many countries is typified by what happened in the United Kingdom, where the main problem with food contamination occurred in upland areas with sheep. Restrictions in the United Kingdom were imposed on sheep in June 1986, once levels had risen to those that caused concern. Initially, the Food and Environment Protection Act, 1985 was used. Under this act, the Minister of Agriculture, Fisheries and Food could issue an order banning the sale and movement for foods contaminated over a specified level, in this case 1,000 Bq/kg (U.K. Government, 1986). These orders can be used for any contaminant.

Other countries in the EU were also affected by the Chernobyl accident and imposed their own restrictions since there are no internationally agreed accident limits. Concern was registered that imports of food from other countries, both inside and outside the EU, may exceed the levels that individual countries set. This led to the European Commission issuing Council Regulation (EURATOM) No. 3954/87, under Articles 31 and 2(b) of the EURATOM Treaty, which lays down the procedure for determining maximum permissible levels (MPLs) of radioactivity in foods immediately following a nuclear accident or incident within the EU. This regulation was initially driven by individual member states' setting their own national limits, with inevitable effects on trade within the community. This regulation also set maximum levels in dairy products and other foods and stated that levels will be subsequently set for baby foods, liquid foods, and minor foods. Further European Council regulations implementing these levels are listed in Exhibit 15-1. Animal feeding stuffs are also covered by these regulations.

In 1990, the European Council issued a regulation (737/90) that specifically concerned the Chernobyl accident. This regulation dealt with food and feeding stuffs originating outside the EU. It set up a similar framework to establish MPLs. It set low MPLs for cesium-134 and cesium-137 in milk and foods intended for feeding infants. It also listed those products that are exempt. This regulation has since been modified, and its scope has been extended to the year 2000.

Exhibit 15–1 EU Legislation on MPLs in Foods, Following a Nuclear Accident

Regulation	Effects
Council regulation (EURATOM) No. 3954/87 (EC, 1987)	Creates framework for laying down derived limits; sets levels in dairy products and other foods; undertaken to lay down levels in other foods.
Commission Regulation (EURATOM) No. 994/89 (EC, 1989a)	Lists minor foods; sets MPLs for minor food.
Council Regulation (EURATOM) No. 2218/89 (EC, 1989b)	Sets MPLs for baby foods and liquid.
Council Regulation (EEC) No. 2219/89 (EC, 1989c)	Describes export conditions for EEC countries following imposition of an MPL; if food or feeding stuff exceeds levels in Articles 2 and 3 of No. 3954/87, it cannot be exported.
Council Regulation (EURATOM) No. 770/90 (EC, 1990a)	Sets MPL for cesium-134 and cesium-137 in feeding stuffs.
Council Regulation (EEC) No. 737/90 (EC, 1990b)	Creates framework for setting levels in human food originating in a Third World (non-EEC) country; sets levels for cesium-134 and cesium-137 in food intended for infants and levels in other food; food that is not covered by this regulation.
Council Regulation (European Community) No. 686/95 (EC, 1995)	Extends expiry date of previous regulation to March 31, 2000.
Commission Regulation (European Community) No. 727/97 (EC, 1997)	Addresses products excluded from Council Regulation 737/90.

Immediately following a nuclear accident, the European Commission is likely to issue regulations that provide a coarse control on potentially hazardous foods while the situation is studied more carefully. The proposed regulations are likely to set MPLs as shown in Table 15–2 and would apply to marketing and exporting. The respective regulation would stay in force for as short a time as possible, but in any case, no longer than three months after which time a group of experts should review the provision of the regulation. The inapplicability of MPLs in normal operations, where discharges are carefully controlled and assessed, should be noted.

In 1993, ICRP updated its advice on public protection in the event of an accident involving radioactivity (ICRP, 1993). These principles stated that the averted

Table 15–2 Maximum Permitted Levels (Bq/kg or Bq/l) in Foods Following a Nuclear Emergency

Isotopes	Baby Food	Dairy Produce	Other Foods	Liquid Foods
Pu-239 and Am-241	1	20	80	20
Sr-90	75	125	750	125
I-131	150	500	2,000	500
All other nuclides with half-lifes <10 days (Cs-134 and Cs-137)*	400	1,000	1,250	1,000

*Not including H-3, C-14, or K-40.
Source: Data from European Union legislation.

individual dose should be optimized and in the range 5 mSv to 50 mSv. The levels proposed by ICRP for the withdrawal of foods (not MPLs in foods) are 1 kBq/kg to 10 kBq/kg for beta-emitting radionuclides and 1 Bq/kg to 100 Bq/kg for alpha-emitting radionuclides. In 1994, the IAEA published its view on the intervention levels for nuclear or radiation emergencies (IAEA, 1994). These levels are given in Table 15–3 (compare with those given by ICRP above). The IAEA levels are the same as those published in 1989 by the CAC (WHO, 1989). These sets of recommendations have been considered when advice is issued by national authorities (NRPB, 1994), but are not included in regional or national regulations.

THE NATIONAL RADIOLOGICAL PROTECTION BOARD

The National Radiological Protection Board (NRPB) is the nondepartmental government body that advises the U.K. government departments and agencies on the scientific and technical background to all aspects of radiation protection, including both ionizing and nonionizing radiation. Advice is also available to the general public on inquiry, and to commercial organizations on a repayment basis. The NRPB was set up under the Radiation Protection Act, 1970, under the auspices of the Department of Health, with a duty to acquire knowledge on radiation protection for humankind and to provide advice to the government of the United Kingdom (U.K. Government, 1970). The strategic direction of the NRPB is provided by the chair and the members of the board who are appointed by, and responsible to, health ministers. Members of the board are nationally recognized experts in various fields of radiation protection, public health, and epidemiology. They also monitor the scientific program of the NRPB as a whole. The board delegates responsibility for the daily management of resources and scientific performance to the director (chief executive). The various areas of radiation protection are split into four divisions: Standards and Services, Environmental and Physical Sciences, Biomedical Sciences, and Nonionizing Radiation. Each of these divisions is headed by an assistant director.

In 1977, the NRPB was directed by the health ministers under the Radiation Protection Act 1970 to give advice on the acceptability in the United Kingdom of

Table 15–3 Comparison of International Intervention Levels for Foods

Organization	Radionuclides	Levels	
		General Consumption	Infant Consumption
ICRP	β/γ emitters	1–10 kBq/kg	
	α emitters	10–100 Bq/kg	
IAEA			
	Cs-134, Cs-137, Ru-103, Ru-106, Sr-89	1	1
	I-131	1	0.1
	Sr-90	0.1	0.1
	Am-241, Pu-238, Pu-239, Pu-240, Pu-242	0.01	0.001

Source: Data from *Principles for Protecting the Public in a Radiological Emergency,* © 1993, ICRP; and *Intervention Criteria in a Nuclear or Radiation Emergency,* © 1994; IAEA.

standards recommended by international and intergovernmental bodies and also to give advice on implementing countermeasures after an accident. From this date, the board has produced advice on these topics. Since 1990, this advice has been published in a special series of publications entitled Documents of the NRPB.

NRPB also conducts research in several areas that are important for radiological protection. Mechanisms of radiation-induced leukemogenesis are being studied in order to obtain a better understanding of the effects of low doses of radiation (see "Radioactivity and Radiation Protection"). In the area of estimating dose, NRPB works with other countries to calculate dose coefficients that are required in order to estimate doses from ingestion of radionuclides. These dose coefficients are included in the BSS issued by the European Commission in 1996.

Following publication of the latest ICRP recommendations on a general system of radiological protection (ICRP, 1991), the board published a statement on the applicability of these recommendations to the United Kingdom (NRPB, 1993), together with more detailed advice in particular areas (NRPB, 1994). For members of the general public, an annual dose limit of 1 mSv for controlled sources was recommended, without the flexibility offered by the ICRP of allowing higher doses in some years, provided that the average over five years does not exceed 1 mSv/yr. The lifetime fatal cancer risk from exposure at 1 mSv/yr is estimated by ICRP at 0.4 percent, which represents an increase of about 1.5 percent on the natural probability of dying from cancer, which is 20 to 25 percent.

The ICRP recommendations also introduced a new concept, the dose constraint, which is used during the optimization of protection (see "Radioactivity and Radiation Protection"). The NRPB has provided advice on how this concept should be applied in the United Kingdom. In this regard, the dose constraint is an upper bound on the annual dose that members of the public may receive from the planned operation of a single controlled source, such as a nuclear power plant. The NRPB recommends an upper value for the dose constraint for members of the public of 0.3 mSv/yr. The advice confirms that while most current nuclear plants should be able to operate within such a constraint, some may not be able to do so. In such cases, the NRPB advises that the regulatory body should ensure that the resulting doses are ALARA and within dose limits. Advice is provided on estimating the dose that is compared with the dose constraint. The constraint applies to the control of current and proposed facilities and, as such, doses from past discharges are not considered. All relevant exposure pathways from the controlled source back to humans should be taken into account. In practice, this means that exposure via an individual foodstuff must be considered in association with all other relevant foodstuffs, whether affected by liquid, solid, or gaseous wastes, together with exposures from other pathways such as inhalation of radionuclides. Similar considerations apply to estimating doses for comparison with the dose limit; in this case, however, doses from all relevant controlled sources should be considered together with additional controlled exposures from past practices or past controlled discharges.

In order to simplify the assessment of doses to the public, the NRPB publishes generalized derived limits (GDLs) (NRPB, 1996). These are reference levels against which results from environmental monitoring can be compared. They are related to the dose limit by a defined model and are calculated such that compliance with them would ensure virtual certainty of compliance with the limit. The NRPB recommends that GDLs are used as a test against which the results of environmental monitoring or a simple assessment can be compared; if a particular measurement exceeds 10 percent of the relevant GDL, more detailed investigations are needed.

The government of the United Kingdom has accepted NRPB advice on the protection of the public from radioactive discharges as published in the white paper on *A Review of Radioactive Waste Management Policy* (U.K. Government, 1995). Here, an additional proviso on the dose constraint for exposure from two contiguous sites or operation was put forward, in that the site target should be 500 μSv and that this would include doses associated with previous discharges. This affects sites where two independent operators are producing discharges affecting the same critical group, with the pro rata source constraint being 250 μSv.

ASSESSMENT OF DOSES FROM CONSUMPTION OF RADIONUCLIDES IN FOODS

As previously stated (see "Radioactivity and Radiation Protection"), approaches to limiting levels of radionuclides in foods differ from those used for other contaminants. No acceptable or permissible levels for radionuclides are specified by any international organization for normal circumstances. Instead the approach used is to consider the dose received by the general public from all anthropogenic sources (including that arising from processes that lead to enhanced levels of natural radionuclides, but excluding medical sources). The dose from consumption of radionuclides in food, from all anthropogenic sources, forms only one component. Limits are specified for this total dose. The limits are, in practice, compared to the doses received by members of the critical group, i.e., that group who, because of their location or habits, are expected to receive the highest doses.

Assessment of the food chain component of the total whole-body dose starts with evaluation of the radioactivity in different foods. Depending on whether the assessment is being carried out in order to judge the acceptability of a proposed facility or to consider an existing situation, the levels of radioactivity can be evaluated either by calculating predicted activities or by measuring levels in samples of food. If predictive calculations are being carried out, consideration needs to be given to the source of the radioactivity—how much of various radionuclides are discharged from the facility in the different waste streams, the exact location of the discharge point, and which chemical and physical forms are appropriate.

Calculations are then made on how the material is transported within the atmosphere or aquatic environment. This uses models such as the Gaussian plume model (Clarke, 1979) or U.K. Atmospheric Dispersion Modelling System (UKADMS) (Carruthers et al., 1994) for atmospheric dispersion, and MAFF Irish Sea Modelling Aid (MIRMAID) (Gurbutt et al., 1992) for aquatic dispersion. It is then necessary to assess the uptake of material by crops and livestock or fish and aquatic foods. Such calculations involve a range of factors and parameters, some of which are not fully quantified. Some recommended data values have been published by the IAEA and other bodies, but these generally have wide ranges of uncertainty. In the assessments carried out by the U.K.'s MAFF, to ascertain whether proposed levels of discharge are acceptable, conservative assumptions are used to ensure that the levels of radioactivity are not underestimated, without producing an unrealistic overestimation. For existing facilities and processes, the modeling and calculation stage can be replaced by sampling crops and livestock, and analyzing radionuclide content.

Once the levels of activity in the foods have been evaluated, the amount of the radionuclides consumed can be evaluated. Generally as a conservative assumption where no data are available, no account is taken of the loss of radioactivity as a consequence of possible food preparation such as peeling and washing although nonedible portions are removed. Where information is available (e.g., from habit surveys), the food processing techniques observed are applied.

The intake of the radionuclides estimated from these calculations is evaluated by multiplying the activity in the foods by amounts of food consumed. Data on the latter are obtained either from generic national data or by using site-specific data. Wherever possible, consumption data for different age groups is used since the behavior of radionuclides within the body, and hence the doses received, varies with age. When considering the doses to the general public resulting from discharges by a nuclear facility, the conservative assumption is made that all foods consumed are produced in the locality wherever possible, thus maximizing the estimate of radioactivity consumed. However, when a survey has been conducted in the locality to identify the foods and amounts consumed by members of the critical group, data from this survey are used. A further conservative assumption is made in using consumption rates typical of a higher than average consumer. This focus on the high-level consumer, thereby ensuring that the level of exposure of the average consumer is acceptable, is consistent with that used for assessing other contaminants in the food chain. Once intakes of the various radionuclides have been evaluated, the dose can be calculated by multiplying by the dose-per-unit-intake coefficients. These coefficients have been calculated by organizations such as the ICRP and national authorities and are derived from studies of the biokinetics of the nuclides in vivo. Separate values have been derived for different age groups; doses from ingestion are evaluated for the separate groups.

In regulating discharges, doses are evaluated for different age groups and then combined with those for other pathways such as inhalation, irradiation from the facility, and irradiation by contaminated sediments to give the total dose. Again these other pathways are evaluated for all age groups, and conservative, but realistic, assumptions are made so that the dose calculated represents an upper-bound value. The dose of the age group receiving the highest dose is then compared with the dose limit.

This methodology has a number of advantages over other approaches. It enables a holistic approach to be taken to a facility or practice, and the total detriment from radioactive discharges can be evaluated. Also, it is possible to evaluate the dose from a wide range of radionuclides, or combinations of radionuclides, on a uniform basis. This is important since discharges of a single nuclide rarely occur; most discharges contain more than one radionuclide.

Because radionuclides generally arise in combination and often in a range of foodstuffs, a holistic approach enables rational assessments for all radionuclides and all pathways. The absence of simple limits on concentrations of radionuclides is sometimes difficult to understand, but such limits have no place in a complex system.

ENFORCEMENT IN THE UNITED KINGDOM

The standards used for contamination of foods by radioactivity are effectively worked out on an individual basis for each case where a source of potential pollution exists. In the United Kingdom, the parts of government responsible for control of these discharges are the Department of the Environment, Transport and the Regions (DETR); the MAFF; the Scottish Parliament; and the National Assembly of Wales. Regulation of discharges is currently undertaken by either the Environment Agency (EA) or the Scottish Environmental Protection Agency.

The Radioactive Substances Act, 1948, sets out general provisions to control radioactive substances (U.K. Government, 1948). Detailed rules governing disposal of radioactive waste were defined in the Atomic Energy Authority Act, 1954 (U.K. Government, 1954). This latter act led to the formation of the Atomic Energy Authority. It gave responsibility for authorizing discharges and consulting interested parties to MAFF and the Ministry of Housing and Local Government (now DETR). The Radioactive Substances Act, 1960, kept the same powers as the 1954 Act, but widened the scope of the sites covered (U.K. Government, 1960). The 1960 act remained in force until 1993 when the Radioactive Substances Act (1993) took account of the changes that had taken place over the previous 33 years (U.K. Government, 1993). This latter act was revised in 1995 by the Environment Act, which led to the formation of the EA (U.K. Government, 1995). The main effect was that MAFF was no longer a coauthorizing department, but became, along with DETR, a statutory consultee for authorizations issued by the EA. Recognizing the importance of the food chain in radiation protection, the Minister of Agriculture has powers of direction over the EA to ensure that the food chain is not compromised by discharges from sites using radioactive materials.

Whenever authorizations are changed at major sites, an individual dose assessment is undertaken. In this way, the individual effects of each radionuclide discharged is assessed and the total impact of all discharges from the site can be seen in the wider context of exposure to individuals and the population as a whole. The individual dose is compared with a dose constraint, set at 300 µSv by the NRPB, for an individual site. This ensures that the dose limit of 1,000 µSv is not approached if individuals are exposed to discharges from more the one site. The idea of dose constraints was put forward by the ICRP in its latest set of recommendations (ICRP, 1991).

The authorities in the United Kingdom measure the effects of individual discharges from the larger (licensed) sites in order to monitor the effects on the food chain. In addition, samples of food are taken throughout the country and in areas distant from nuclear sites to assess the overall effect of all discharges on the food chain. These results are routinely reported by MAFF (MAFF, 1996).

CONCLUSION

In the United Kingdom, radioactivity can be measured at very low levels in the food chain, and, therefore, very small increments of risk to human health can be assessed.

In the early days, ad hoc groups recommended controls on exposure, for example the International X-ray and Radium Protection Committee and, latterly, the ICRP. As international relations have developed, with nuclear weapons being produced and nuclear power generated, recommendations for control of radiation exposure and hence health risk have been fitted into the developing international fora—as witnessed by the setting up of the IAEA.

The concepts of radiation protection and related risk management are now set on an international basis. They are promulgated through international and regional bodies for incorporation into national legislation and, in emergency situations, derived limits have been set to expedite response. Levels of individual radioactive contaminants are not set against an international benchmark, but are derived from the local conditions that pertain in individual nations. Narrow limits are not set on individual radionuclides, but instead are set on the associated increase in risk to health.

REFERENCES

Arnold, L. (1995). *Windscale, 1957: Anatomy of a Nuclear Accident.* London: Chapman and Hall.

Bertell, R. (1985). *No Immediate Danger, Prognosis for a Radioactive Earth.* London: The Women's Press.

Carruthers, D. J., Holroyd, R. J., Hunt, J. C. R., Weng, W. S., Robins, A. G., Apsley, D. D., Thompson, D. J. & Smith, F. B. (1994). A new approach to modelling dispersion in the earth's atmospheric boundary layer. *J Wind Eng Ind Aerodynamics* 52, 139–153.

Caufield, C. (1989). *Multiple Exposures.* London: Martin Secker & Warburg.

Clarke, R. H. (1979). A Model for the Short and Medium Range Dispersion of Radionuclides Released to the Atmosphere. R-91. Chilton, England: National Radiological Protection Board.

Edwards, R. (1997). Radiation roulette. *New Scientist* 2103.

European Community. (1976). Council directive 76/579/EURATOM laying down the revised safety standards for the health protection of the general public and workers against the dangers of ionising radiation. *Off J Eur Communities* L187, 1–14.

European Community. (1979). Council directive 79/343/EURATOM amending directive 76/579/EURATOM laying down the revised basic safety standards for the health protection of the general public and workers against the dangers of ionising radiation. *Off J Eur Communities* L83, 18.

European Community. (1980). Council directive 80/836/EURATOM amending the directives laying down the basic safety standards for the health protection of the general public and workers against the dangers of ionising radiation. *Off J Eur Communities* L246, 1–14.

European Community. (1984). Council directive 84/467/EURATOM amending directive 80/836/EURATOM as regards the basic safety standards for the health protection of the general public and workers against the dangers of ionising radiation. *Off J Eur Communities* L265, 4.

European Community. (1987). Council Regulation (EURATOM) No. 3954/87 of 2 December 1987 laying down maximum permitted levels of radioactive contamination of foodstuffs and feedingstuffs following a nuclear accident or any other radiological emergency. *Off. J. Eur. Community* L371, 11.

European Community. (1989a). Council Regulation (EURATOM) No. 944/89 of 12 April 1989 laying down maximum permitted levels of radioactive contamination in minor foodstuffs following a nuclear accident or any other case of radiological emergency. *Off. J. Eur. Community* L101, 17.

European Community. (1989b). Council Regulation (EURATOM) No. 2218/89 of 18 July 1989 amending Regulation (EURATOM) No. 3954/87 laying down the maximum permitted levels of radioactive contamination of foodstuffs and feedingstuffs following a nuclear accident or any other case of radiological emergency. *Off. J. Eur. Community* L211, 1.

European Community. (1989c). Council Regulation (EEC) No. 2219/89 of 18 July 1989 on the special conditions for exporting foodstuffs and feedingstuffs following a nuclear accident or any other case of radiological emergency. *Off. J. Eur. Community* L221, 4.

European Community. (1990a). Council Regulation (EEC) No. 737/90 of 22 March 1990 on the conditions governing imports of agricultural products originating in third world countries following the accident at Chernobyl nuclear power-station. *Off. J. Eur. Community* L082, 1.

European Community. (1990b). Commission Regulation (EURATOM) No. 770/90 of 29 March 1990 laying down maximum permitted levels of radioactive contamination of feedingstuffs following a nuclear accident or other case of radiological emergency. *Off. J. Eur. Community* L083/90.

European Community. (1995). Council Regulation (EC) No. 686/95 of 28 March 1995 Extending Regulation (EEC) No. 737/90 on the conditions governing imports of agricultural products originating in third world countries following the accident at the Chernobyl nuclear power-station. *Off. J. Eur. Community* L071, 15.

European Community. (1996). Council directive 96/29/EURATOM laying down the basic safety standards for the protection of the health of workers and the general public against the dangers arising from radiation. *Off J Eur Community* L159, 1–114.

European Community. (1997). Commission Regulation (EC) No. 727/97 of 24 April 1997 establishing a list of products excluded from the application of Council Regulation (EEC) No. 737/90 on the conditions governing imports of agricultural products originating from third world countries following the accident at the Chernobyl nuclear power station. *Off. J. Eur. Community* L108, 16–18.

Gurbutt, P. A., Kershaw, P. J., Penthreath, R. J., Woodhead, D. S., Durance, J. A., Camplin, W. C. & Austin, L. S. (1992). *MIRMAID: The MAFF Irish Sea Model*. Fisheries Research Technical Report. Lowestoft, England: MAFF Directorate Fisheries Research.

Hughes, J. S. and O'Riordan, M. C. (1993). *Radiation Exposure of the UK Population—1993 Review*. R-263. Oxon, United Kingdom: National Radiological Protection Board.

Hunt, G. J., Leonard, D. R. P. & Lovett, M. B. (1990). Transfer of environmental plutonium and americium across the human gut: a second study. *Sci Total Environ* 90, 273–282.

International Atomic Energy Agency. (1960). *The Agency's Health and Safety Measures*. INFCIRC/18. Vienna: International Atomic Energy Agency.

International Atomic Energy Agency. (1967). *Basic Safety Standards for Radiation Protection*. Safety Series, No. 9. IAEA, Vienna: International Atomic Energy Agency.

International Atomic Energy Agency. (1977). *Recommendations of the International Commission on Radiological Protection*. ICRP Publication 26, Annals of the ICRP 1. Vienna: International Atomic Energy Agency.

International Atomic Energy Agency. (1994). *Intervention Criteria in a Nuclear or Radiation Emergency.* Safety Series, No. 109. Vienna: International Atomic Energy Agency.

International Atomic Energy Agency. (1995). *Environmental Impact of Radioactive Releases.* Proceedings of a Symposium, Vienna, May 8–12, 1995. Vienna: International Atomic Energy Agency.

International Atomic Energy Agency. (1996). *International Basic Safety Standards for Protection against Ionising Radiation and for the Safety of Radiation Sources.* Safety Series, No. 115. Vienna: International Atomic Energy Agency.

International Commission on Radiological Protection. (1958). *Recommendations of the International Commission on Radiological Protection.* 9 September 1958. London: Pergamon Press.

International Commission on Radiological Protection. (1959). *Recommendations of the International Commission on Radiological Protection.* ICRP Publication 1. Oxford: Pergamon Press.

International Commission on Radiological Protection. (1977). *Recommendations of the International Commission on Radiological Protection.* Publication 26. Annals of the ICRP 1(3). Oxford: Pergamon Press.

International Commission on Radiological Protection. (1985). Statement from the 1985 Paris meeting of the International Commission on Radiological Protection. Annals of the ICRP 15(3). Oxford: Pergamon Press.

International Commission on Radiological Protection. (1991). *Recommendations of the International Commission on Radiological Protection.* ICRP Publication 60, Annals of the ICRP 21. Oxford: Pergamon Press.

International Commission on Radiological Protection. (1993). *Principles for the Intervention for the Protection of the Public in a Radiological Emergency.* ICRP Publication 63, Annals of the ICRP 22. Vienna: International Atomic Energy Agency.

International Commission on Radiological Protection. (1995). *Age-Dependent Doses to Members of the Public from Intake of Radionuclides: Part 5 Compilation of Ingestion and Inhalation Dose Coefficients.* ICRP Publication 72. Annals of the ICRP 26. Vienna: International Atomic Energy Agency.

Martin, A. & Harbison, S. A. (1996). *Introduction to Radiation Protection.* London: Chapman and Hall.

Ministry of Agriculture, Fisheries and Food. (1996). *Radioactivity in Food and the Environment, 1995.* London: Ministry of Agriculture, Fisheries and Food.

National Radiological Protection Board. (1993). *Occupational, Public and Medical Exposure.* Documents of the NRPB 4, no. 2. Clinton, England: National Radiological Protection Board.

National Radiological Protection Board. (1994). *Guidance on Restrictions on Food and Water Following a Radiological Accident.* Documents of the NRPB 5, no. 1. Clinton, England: National Radiological Protection Board.

National Radiological Protection Board. (1996). Generalised Derived Limits for Radioisotopes of Strontium, Ruthenium, Iodine, Caesium, Plutonium, Americium and Curium. Documents of the NRPB 7, No. 1. Clinton, England: National Radiological Protection Board.

Union of International Associations. (1994). *Yearbook of International Associations, 1994/5.* Munich, Germany: K.G. Saur.

United Kingdom Government. (1948). The Radioactive Substances Act 1948 (c.37). London: Her Majesty's Stationary Office.

United Kingdom Government. (1954). *Atomic Energy Authority Act. 1954* (c.32). London: Her Majesty's Stationary Office.

United Kingdom Government. (1960). The Radioactive Substances Act 1960 (c.34). London: Her Majesty's Stationary Office.

United Kingdom Government. (1970). The Radiation Protection Act 1970 (c.46). London: Her Majesty's Stationary Office.

United Kingdom Government. (1972). Treaty Establishing European Atomic Energy Community. Rome, March 25, 1957. Cmnd. 4865. London: Her Majesty's Stationary Office.

United Kingdom Government. (1986). The Food Protection (Emergency Prohibitions) Order 1986. SI 1986/1027. London: Her Majesty's Stationary Office.

United Kingdom Government. (1993). The Radioactive Substances Act 1993 (c.12). London: Her Majesty's Stationary Office.

United Kingdom Government. (1995). *A Review of Radioactive Waste Management Policy*. Cmnd 2919. London: Her Majesty's Stationary Office.

World Health Organization. (1989). Supplement 1 to *Codex Alimentarius*, Vol. XVII. Vienna: World Health Organization.

Chapter 16

Establishment of Codex Microbiological Criteria for Foods

Mike van Schothorst and Anthony Baird-Parker

INTRODUCTION

Most food-borne diseases, if not all, can be prevented by applying the basic principles of food hygiene throughout the food chain, from the primary producer to the final consumer. The primary responsibility of producing, manufacturing, and selling safe food is, of course, in the hands of the industry engaged in these activities. This is clearly expressed in the Codex Alimentarius Commission (CAC) document on *General Principles of Food Hygiene* (CAC, 1997a). The consumer's role is reflected in the definition of food safety: assurance that food will not cause harm to the consumer when it is prepared and/or eaten according to its intended use.

Governmental agencies also have an important role to play in ensuring that food is safe and suitable for consumption, as well as in maintaining confidence in internationally traded food. Three principal means are used by governments in food control: (1) inspection of facilities and operations, (2) testing of food samples, and (3) education. Education and training of food handlers and consumers help them to understand and apply safe food-handling practices. Inspection of facilities is intended to provide evidence that operations were designed and practices are applied to ensure that safe food is consistently delivered, and the results of testing should confirm this. Microbiological criteria, used in testing, should distinguish between an acceptable and unacceptable product or between acceptable and unacceptable food-processing and food-handling practices.

To ensure the safety of imported food, governments have in the past placed much emphasis on microbiological testing of incoming consignments. This situation is now changing, partly as a consequence of the signing of the Application of Sanitary and Phytosanitary Measures (SPS Agreement). The implementation of the SPS Agreement is intended to facilitate the free movement of foods across borders, by ensuring that measures established by countries to protect human health are scientifically justified, and are not used as nontariff barriers to trade in foods (WTO, 1995). In principle, this should reduce the amount of testing because microbiological examination of consignments is not able to provide the scientifically justified evidence that food safety was ensured by the producer in the country of origin, as will be discussed below.

To facilitate trade of safe food, emphasis is now on transparency and equivalence of food control systems. Thus the exporting country is responsible for providing the importing country with all the information necessary to judge the safety and suitability of food for consumption by the local population. Codex has developed a number of standards, codes, and guidelines that are necessary to achieve this transparency and equivalence. Particularly important documents are the *General Principles of Food Hygiene* (CAC, 1997a) and its Annex entitled, *Hazard Analysis and Critical Control Point System and Guidelines for Its Application*. It is very important to understand how Codex approaches the subject of microbiological criteria. The development and application of such criteria are described in the *Principles for the Establishment and Application of Microbiological Criteria for Foods* (CAC, 1997b).

This chapter deals with some essential elements of establishing and applying microbiological criteria, and how results should be interpreted and used.

HISTORY

After it was recognized that microbes could cause food-borne diseases, notions such as "pathogens should be absent" have been used in food regulations. In Codex codes of practice the wording was changed in the early 1970s:

> When tested by appropriate methods of sampling and examination, the product:
> - shall be free from microorganisms in amounts which may represent a hazard to health
> - shall be free from parasites which may represent a hazard to health
> - shall not contain any substance originating from microorganisms in amounts which may represent a hazard to health.

This wording was partially based on the work of the International Commission on Microbiological Specifications for Foods (ICMSF, 1986). The ICMSF was formed in 1962 in response to the need for internationally acceptable and authoritative decisions on microbiological limits for foods moving in international trade. Throughout the years, it has provided expertise to the Codex Committee on Food Hygiene (CCFH) by members' participating in regular CCFH meetings, Food and Agriculture Organization/World Health Organization (FAO/WHO) expert consultations, and working groups, and by writing papers for discussion.

The first microbiological criteria for Codex were developed during a Joint FAO/WHO Expert Consultation, in spring 1976, for egg products (Christian, 1983). The criteria were based on ICMSF sampling plans (ICMSF, 1986) and International Standards Organization methods of examination. During the consultation, it became apparent that no guidelines existed for the establishment of microbiological criteria and that these were essential for the standardization and harmonization activities of the Codex commodity committees. Consequently, the main topic on the agenda of the second Joint FAO/WHO Expert Consultation, in 1977, was the elaboration of *Principles for the Establishment and Application of Microbiological Criteria for Foods*. The text was accepted with a few amendments by the CCFH and adopted by the CAC in 1981 (CAC, 1981). It was issued first as a single document and later, in 1986, included in the *Procedural Manual* (CAC, 1986). In the early 1990s, the hazard analysis critical control point (HACCP) concept was adopted and the *General Principles of Food Hygiene* (CAC, 1997a) was being revised.

As a consequence of these developments, it was felt that the criteria document also needed to be revised. At the request of the CCFH, the ICMSF prepared a first draft, which was finalized by the CCFH committee in 1996 and adopted by the CAC in 1997. It is this document, now called *Principles for the Establishment and Application of Microbiological Criteria for Foods* (CAC, 1997b), that provides the basis of the rest of this chapter. The importance of this document is reflected in the recommendation that in commodity standards the following text should be used: "The products should comply with any microbiological criteria established in accordance with the *Principles for the Establishment and Application of Microbiological Criteria for Foods* (CAC/GL 21–1997)" (CAC, 1997b).

ESTABLISHMENT OF MICROBIOLOGICAL CRITERIA

General Considerations

According to the Codex document, *Principles for the Establishment and Application of Microbiological Criteria for Foods* (CAC, 1997b), a microbiological criterion for food defines the acceptability of a product or a food lot, based on the absence or presence, or number of microorganisms, and/or quantity of their toxins/metabolites, per unit(s) of mass, volume, area, or lot. The text mentions that a microbiological criterion should be based on need (principally on epidemiological evidence of a health concern) and applicability (utility and particularly a positive health benefit) and should contain:

- limits for each microorganism or toxin of concern in a specific food
- a sampling plan for each of these
- methods of analysis
- designation of place in the food chain where the criterion should apply
- action(s) to be taken if a food lot failed to meet the criterion

The principles are intended to "give guidance on the establishment and application of microbiological criteria for foods at any point of the food chain from primary producer to final consumption" (CAC, 1997b). Taking this into consideration and the various uses of criteria, it is clear that before a criterion is elaborated at least three questions should be answered:

1. What is the purpose of the criterion (i.e., is there a need)?
2. Where is the criterion to be applied?
3. What are the consequences when the criterion is not met?

If there is no clearly identified need, a criterion should not be developed. Such need is demonstrated, for example, "by epidemiological evidence that the food under consideration may represent a public health risk and that the criterion is meaningful for consumer protection or as the result of risk assessment"(CAC, 1997b). The second part of this sentence refers to the SPS Agreement, which states that risk assessment should be used to provide the scientific basis for national food regulations on food safety.

Criteria to be applied at a food production or preparation site may differ from the ones applied at the retail level or at a port of entry into a country. This is reflected in the phrase that to fulfil the purposes of a criterion, consideration should be given to "the likelihood and consequences of microbiological contamination and/or growth during (subsequent) handling storage and use" (CAC, 1997b).

The consequences of not meeting a criterion may vary depending on the purpose of the testing and whether the criterion is voluntary (applied by the industry) or mandatory. Whatever the action to be taken might be, it should be decided upon before developing a criterion, and not after applying a criterion.

Other aspects to be considered in setting microbiological criteria include:

- the microbiological status of the raw material(s)
- the effect of processing
- the intended use of the food
- the category(ies) of consumers concerned
- the cost/benefit ratio associated with applying the criterion

The first two aspects speak for themselves. If raw beef contains inevitable low numbers of *Listeria monocytogenes,* and when the process does not include a killing step, it is unrealistic to specify that the ground beef patty should not contain this microorganism in x number of samples of 25 g. The Codex text clearly states that the criterion "should be technically attainable by applying Good Manufacturing Practices" (CAC, 1997b). If a product is fermented, it is useless to set a criterion on aerobic plate counts, for example.

The intended use is an important consideration. The Codex definition for food safety is, "assurance that food will not cause harm to the consumer when it is prepared and/or eaten according to its intended use" (CAC, 1997b). A frozen pizza that needs to be thoroughly heated before consumption could contain an odd *Salmonella* without causing harm to the consumer. A stringent sampling plan for *Salmonella* in such a food would normally make no sense; proper labeling with adequate instructions for preparation and use would be more effective.

Consideration of the intended use should also include who is going to use the product: a professional caterer or a homemaker. Even more important is which group of consumers is targeted. For example, babies, ill people, the elderly, and immunosuppressed individuals are more vulnerable than healthy athletes. Thus more care should be taken in producing or preparing food for these categories of consumers. Stringency of sampling plans and their limits should reflect this.

Lastly, the cost/benefit considerations should focus on whether setting and applying or enforcing a criterion would be an effective means of using the available resources. This type of consideration falls under the heading of risk management

(FAO/WHO, 1997), and particularly, "... choosing an adequate Risk Management Option" (Section 11.3.1).

Microbiological Aspects of Criteria

The Codex documents pertain to:

- bacteria, viruses, yeasts, molds, and algae
- parasitic protozoa and helminths
- their toxins/metabolites

The microorganisms included in a criterion should be widely accepted as relevant to the particular food and technology—as pathogens, as indicator organisms, or as spoilage organisms. Fecal streptococci, even if they are mentioned in criteria for certain foods, would at present not fall in this category. When criteria include indicator organisms, it should be clear what their presence or numbers indicate (e.g., unacceptable survival of specific pathogens and/or unacceptable (re)contamination or multiplication). When an indicator test is used, it should be stated "whether it is used to indicate unsatisfactory hygienic practices or a health hazard" (CAC, 1997b).

The mere finding of a pathogen does not necessarily indicate a threat to public health. The above example of *Salmonella* in a frozen pizza illustrates this point; the Codex text (CAC, 1997b) mentions *Clostridium perfringens*, *Staphylococcus aureus*, and *Vibrio parahaemolyticus* as examples.

The quite elaborate methods section of the Codex document stresses that only reliable, if possible validated and standardized, methods are used. Rapid tests should preferably be used for the examination of perishable food.

Limits

Whether foods are acceptable or not is defined in a criterion by the

- number of samples examined
- size of the analytical unit
- limits
- number of units that should conform to these limits

A limit can be a maximum number of an indicator organism, but also absence of a pathogen in an analytical unit. Limits should be based on data gathered at various production establishments operating under good manufacturing practices (GMP) and applying HACCP analysis because they should be meaningful and applicable to a variety of similar products. They should "take into consideration the risk associated with the microorganisms, and the conditions under which the food is (intended and) expected to be handled and consumed" (CAC, 1997b). They should also "take account of the likelihood of uneven distribution of microorganisms in the food and the inherent variability of the analytical procedure" (CAC, 1997b).

Sampling Plans

A sampling plan includes the sampling procedure and the decision criteria to be applied to a food lot, and is based on examination of a prescribed number of analytical sample units by defined methods. Sampling plans should be administratively and economically feasible. A sampling plan may define the probability of detecting a microorganism (or group of microorganisms) in a food or that a specified concentration of microorganisms is not exceeded. It must be recognized that no practical sampling plan can ensure absence of the target microorganism and that the concentration of microorganisms measured may be exceeded in a part of the lot that was not sampled.

The probability of acceptance of a food lot with a concentration of microorganisms above that specified in the sampling plan is termed the consumer risk. The rejection of a food with a level of microorganisms below the acceptable level is called the producer risk. Both types of risk must always be considered when setting a sampling plan.

According to the Codex (CAC, 1997b), sampling plans should take into account:

- risks to public health associated with the hazard (severity and likelihood of occurrence)
- the susceptibility of the target group of consumers (very young, very old, or immunocompromised)
- the heterogeneity of distribution of microorganisms or the randomness of sampling
- the acceptable quality level (percentage of nonconforming or defective sample units tolerated) and the desired statistical probability of accepting or rejecting a nonconforming lot

The Codex document does not describe how these sampling plans have to be developed, but makes reference to the two- and three-class attribute plans published by the ICMSF (ICMSF, 1986). The ICMSF approach distinguished three categories of hazards based upon the relative degree of severity:

1. severe hazards (e.g., *Clostridium botulinum*, verotoxinogenic *Escherichia coli*, and *Salmonella typhi*)
2. moderate hazards, potentially extensive spread (e.g., *Shigella*, *Salmonella*, and enterotoxigenic *E. coli*)
3. moderate hazards, limited spread (e.g., *Staphylococcus aureus*, *Clostridium perfringens*, and *Bacillus cereus*)

This categorization and the examples were based on the best epidemiological data available at the time of publication. The categories may need to be revised as a result of new risk assessment procedures.

The other factor to be considered is the likelihood of occurrence of the hazard (i.e., risk), taking into account the anticipated conditions of use. Here the ICMSF (1986) again recognized three categories:

1. situations where the risk would decrease
2. situations where the risk would increase
3. situations where the risk would remain the same

Combining the three levels of severity with the categories of likelihood of occurrence leads to different levels of concern called cases by the ICMSF (1986), case 7 being of lowest concern to food safety and case 15 of the highest. This is a basic form of risk assessment. In situations where the likelihood of occurrence of pathogens is reduced before consumption (e.g., by cooking during preparation), cases 7, 10, and 13 apply, depending on the severity of the hazard (Table 16–1). Cases 8, 11, and 14 refer to situations where the likelihood of occurrence would remain the same between the time of sampling and the time of consumption (i.e., where pathogens are unable to multiply in the food under expected conditions of handling, storage, preparation, and use). Cases 9, 12, and 15 refer to situations where multiplication may occur. Based on these nine cases, the ICMSF developed two-class sampling plans in which "n" indicates the number of sample units to be tested and "c" the number of defective sample units that can be accepted. These sampling plans are summarized in Table 16–1. The plans direct more of the available resources for analysis toward those situations with a high level of concern.

In a lot with a given percentage of defective units, the number of sample units examined determines the probability of detecting food lots that are contaminated. The limitation of sampling is that it is neither practical nor cost-effective to attempt to detect, with a high degree of confidence, low levels of contamination in many foods. In particular, it does not work well with many raw foods (e.g., produce), which have been the source of significant problems lately.

It must be realized that only positive results are meaningful, while negative results provide the level of confidence set by the number of sample units tested, assuming

Table 16–1 Plan Stringency (Case) in Relation to Degree of Health Hazard and Conditions

Type of Hazard	Conditions in Which Food Is Expected To Be Handled and Consumed after Sampling in the Usual Course of Events		
	Reduce Degree of Concern	Cause No Change in Concern	May Increase Concern
Health hazard moderate, direct, limited spread	Case 7 n = 5, c = 2	Case 8 n = 5, c = 1	Case 9 n = 10, c = 1
Health hazard moderate, direct, potentially extensive spread	Case 10 n = 5, c = 0	Case 11 n = 10, c = 0	Case 12 n = 20, c = 0
Health hazard severe, direct	Case 13 n = 15, c = 0	Case 14 n = 30, c = 0	Case 15 n = 60, c = 0

Note: n indicates the number of sample units tested; c indicates the number of defective sample units that can be accepted.

Source: Reprinted with permission from *Microorganisms in Foods 2, Sampling for Microbiological Analysis: Principles and Specific Applications*, 2nd ed., © 1986, International Commission on Microbiological Specifications for Foods.

that there is a homogeneous distribution of the pathogen in the lot or that the samples were randomly taken. For example, finding no defectives after testing five sample units gives 95 percent confidence that a lot is less than 50 percent contaminated. Finding no defectives in 30 samples indicates that the lot is less than 10 percent contaminated; and no defectives in 300 samples means that the lot is less than 1 percent contaminated (all at a 95 percent confidence level). Figure 16–1 gives the probabilities of acceptance for three levels of contamination when using cases 10 to 15 and clearly indicates the limitations of using microbiological testing of samples to ensure food safety or to verify the effective implementation of GMP and HACCP.

In order to ensure that the interpretation of the results of testing takes these statistical facts into account, the Codex prescribes that "the statistical performance characteristics or operating characteristics curve should be provided in the sampling plan" (CAC, 1997b).

PURPOSES AND APPLICATIONS OF MICROBIOLOGICAL CRITERIA

Purposes

It was recognized in developing the first edition of the *Principles for the Establishment and Application of Microbiological Criteria for Foods* that criteria intended to assist a producer to better control production could contain requirements that would not be appropriate for compliance testing of a food in international trade. Thus criteria intended to be used for internal control purposes by a food producer could include

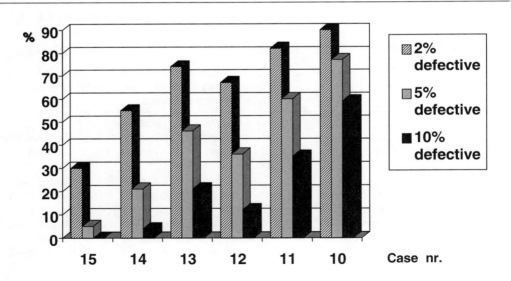

Figure 16–1 Probabilities of Acceptance of Lots with Different Levels of Defective Units. *Source:* Data from *Microorganisms in Foods 2, Sampling for Microbiological Analysis: Principles and Specific Applications*, 2nd ed., © 1986, International Commission on Microbiological Specifications for Foods.

tighter microbiological limits and greater use of indicator organisms. These differences were recognized by the establishment of three types of microbiological criteria:

1. microbiological standards, which were mandatory when incorporated into a Codex standard and adopted in law, and would usually only contain limits for pathogenic microorganisms and toxins of specific concern in that food
2. microbiological specifications, which were advisory and a measure that a food had been produced according to the hygiene requirements in a Codex code of hygienic practice, and could contain limits for a wider range of microorganisms (including general and indicator organisms)
3. microbiological guidelines, which were to be used by the food producer for internal control purposes and which could contain additional requirements to those contained in a standard or specification

In the recent revision of these principles, the separation of criteria into these three types has been replaced by their separation into two types: (1) those that are mandatory and intended to be used by regulatory authorities for official control purposes and (2) those that are intended to be used by industry for internal control and not as such for regulatory purposes. This separation reflects the changing nature of the use of microbiological criteria by industry and enforcement authorities.

It is now widely accepted that HACCP, or HACCP-based systems, are the principal means of ensuring that a food is microbiologically safe, and that microbiological criteria generally provide much less assurance of safety (and quality). Codex (CAC, 1997b) thus recommends that microbiological criteria are used only for control purposes when there is no, or only limited, information that hygienic practices, such as those contained in codes of hygienic or together with HACCP have been applied during the production of the food. However, microbiological criteria, and particularly the limits they contain, have a useful reference function, particularly when used by industry. Thus, they may be used in buying specifications, and within HACCP to provide guidance for setting critical limits for critical control points, and for establishing performance and process criteria. They may also be used as part of the verification process of an HACCP plan (ICMSF, 1988).

The Codex principles (CAC, 1997b) mention that "microbiological criteria may be used to formulate design requirements and to indicate the required microbiological status of raw materials, ingredients, and end-products at any stage of the food chain." Furthermore, "criteria may be relevant to the examination of foods, including raw materials and ingredients, of unknown or uncertain origin or when other means of verifying the efficacy of HACCP-based systems and Good Hygienic Practices are not available."

Application of a Criterion by a Regulatory Authority

A microbiological criterion may be used by the regulatory authorities for border or internal control purposes. Internally, it may also be used to advise food producers of what authorities expect a product to comply with in the marketplace or at point of manufacture, and may be used to audit a producer using the criterion as a measure of product/process acceptability.

If incorporated in law, a criterion may be used to reject noncomplying product. Rejection may lead to sorting, reprocessing, destruction, and/or further investigation to determine appropriate action to be taken, depending on consumer risk. It should be mentioned here that in the context of the SPS Agreement, criteria that are not established according to the Codex document can be challenged by exporting countries or other stakeholders. Moreover, in the World Trade Organization context, microbiological specifications incorporated in Codex codes of practice or standards are the reference criteria for foods in international trade. An example of such a microbiological criterion is presented in Table 16–2.

A regulatory authority relying on product testing to provide assurance that a particular food is safe to eat should be aware that microbiological criteria give poor assurance of safety. Figure 16–1 shows the probability of acceptance of a food subjected to testing using a criterion based on a two-class attribute sampling plan of $n = 5$ to 60 sample units; $c = 0$; $m = 0$. It shows that even extensive testing gives little chance of detecting a low level of microbial contamination. For this reason, Codex (CAC, 1997b) recommends that microbiological criteria should be used only as the basis for acceptance of a food when no other control option is available.

Table 16–2 Microbiological Specifications for Foods for Special Dietary Uses Including Foods for Infants and Children

	Test	Case	Class Plan	n	c	m	M
Dried and Instant Products[a]	Mesophilic Aerobic Bacteria	6	3	5	2[b]	10^3	10^4
	Coliforms	6	3	5	1	<3	20
	Salmonella	12	2	60[c]	0	0[d]	–
Dried Products To Be Heated	Mesophilic Aerobic Bacteria	4	3	5	3[e]	10^4	10^5
	Coliforms	4	3	5	2	10	100
	Salmonella	10	2	5	0	0[d]	–

[a]Including infant formula.
[b]Should have been 1 according to ICMSF sampling plan.
[c]Number increased from 20 to 60 because of sensitive consumers (infants).
[d]Should have been 2 according to ICMSF sampling data.
[e]Analytical units of 25 g.

Note: These specifications were at the time of publication (1982) meant to be "of an advisory nature." They were "intended to increase the assurance that the provisions of hygienic significance have been met, but should not be regarded as mandatory" (CAC, 1982).

Source: Reprinted with permission from *Codex Standards for Foods for Special Dietary Uses, Including Foods for Infants and Children*, CAC/Vol. IX, Ed., 1, © 1982, Food and Agriculture Organization.

Application of a Criterion by a Food Business Operator

Food producers and handlers use criteria to check compliance with regulatory provisions. Such criteria are specific for the product and the stage in the food chain at which they apply. They may be stricter (producer's risk) than the ones used for regulatory purposes, "and should, as such, not be used for legal action" (CAC, 1997b).

In addition to this, "criteria may be applied to formulate product and process design requirements" or, more specifically, process and performance criteria, necessary to meet Food Safety Objectives (Van Schothorst, 1998), a subject outside the scope of this chapter.

The Codex document (CAC, 1997b) further mentions that "criteria may be applied to examine end-products as one of the measures to verify and/or validate the efficacy of a HACCP plan," but not for monitoring purposes. The use of microbiological testing as one of the procedures of verifying HACCP has been proposed by a number of bodies (ICMSF, 1988 and NACMCF, 1992). More recently, the use of such testing has been questioned on the grounds that the confidence of decisions based on such testing is generally poor (Figure 16–1). Assuming that a process were operated to ensure that *Salmonella* would be absent in at least 0.1 percent of production units of a food lot, 3,000 units would need to be examined to have a 95 percent chance of detecting a higher contamination level (ICMSF, 1986). Thus the use of the type of sampling plan that might be applied in a criterion provides little assurance of safety and for the purpose of verification of HACCP; "microbiological testing for pathogens may well be analogous to checking the time recorded on a stop watch by a sundial" (Christian, 1996). This does not mean that there is no use for criteria for verification purposes, but they should be specifically established, and the results carefully interpreted.

CONCLUSION

Microbiological safety criteria are used as tools to protect the health of the consumer. Testing foods for pathogens is, however, not very effective because negative results do not guarantee that the food is indeed safe to eat. Safety is obtained by the application of good hygienic practices and HACCP as a food safety management tool throughout the food chain. However, an importing country may not know whether incoming consignments of foods were indeed produced under good hygienic and safety conditions, and consequently may want to examine the lot. The microbiological criteria to be used in such a situation should be established according to Codex, in order to be in line with the SPS Agreement.

REFERENCES

Christian, J. H. B. (1983). *Microbiological Criteria for Foods. Summary of Recommendations of FAO/WHO Expert Consultations and Working Groups 1975–1981.* VPH/83.54. Geneva: World Health Organization.

Christian, J. H. B. (1996). Microbiological criteria and HACCP: in concert or in conflict? *Int Food Safety News* 4, 4–5.

Codex Alimentarius Commission. (1981*). Principles for the Establishment and Application of Microbiological Criteria for Foods*. ALINORM 81/13 Appendix II. Rome: Codex Alimentarius Commission.

Codex Alimentarius Commission. (1982). *Codex Standards for Foods for Special Dietary Uses, Including Foods for Infants and Children*, 1st ed. Vol. IX. Rome: Codex Alimentarius Commission.

Codex Alimentarius Commission. (1986). *Codex Alimentarius Commission Procedural Manual*. Rome: Codex Alimentarius Commission.

Codex Alimentarius Commission. (1997a). *General Principles of Food Hygiene*. Rome: Codex Alimentarius Commission.

Codex Alimentarius Commission. (1997b). *Principles for the Establishment and Application of Microbiological Criteria for Food*. CAC/GL 21. Rome: Codex Alimentarius Commission.

Food and Agriculture Organization/World Health Organization. (1997). *Risk Management and Food Safety*. Report of a Joint FAO/WHO Expert Consultation, Rome, Italy, January 27–31, 1997. FAO Food and Nutrition Paper 65. Rome: Food and Agriculture Organization/World Health Organization.

International Commission on Microbiological Specifications for Foods. (1986). *Microorganisms in Foods 2, Sampling for Microbiological Analysis: Principles and Specific Applications*, 2nd ed. Toronto, Canada: University of Toronto Press.

International Commission on Microbiological Specifications for Foods. (1988). *Microorganisms in Foods 4, Application of the Hazard Analysis Critical Control Point (HACCP) System To Ensure Microbiological Safety and Quality*. London: Blackwell Scientific Publications.

National Advisory Committee on Microbiological Criteria for Foods. (1992). Hazard analysis critical control point (HACCP) system. *Int J Food Microbiol* 16, 1–23.

Van Schothorst, M. (1998). Principles for the establishment of microbiological food safety objectives and related control measures. In *Food Control*, 6, 379–384.

World Trade Organization. (1995). *Trading into the Future*. Geneva, Switzerland: World Trade Organization.

INDEX

A

Acceptable daily intake (ADI), pesticide residue standard (Codex), 217, 219
Advertising, regulation of, 31
Agreement on the Application of Sanitary and Phytosanitary Standards (SPS Agreement), 6, 65, 94, 257
Agreement on Technical Barriers to Trade (TBT Agreement). *See* TBT Agreement
Application of Risk Analysis to Food Standards Issues, 108
Approved Food Standards and Approved Food Additives, 58
Australia
   food regulation, 57–59
   Food Standards Code, 63
Australia New Zealand (ANZ) Agreement, 59–60, 62
   review of Australian code, 63
Australia New Zealand Food Authority (ANZFA), 59–62
Australia New Zealand Food Authority Act, 59–61, 63
   objectives of, 61–62, 65
   functions of, 60
Australia New Zealand Food Identification System (ANZFIS), 72
Australia/New Zealand food regulation, 59–76
   amending food standards, 61
   contaminant regulation, 72–76
   food additive regulation, 66–72
   food identification system, 71–72
   food standards code, 60
   joint standards proposal, 70–71
   legislation, 57–60
   proposed food standards code, 63–65
   risk assessment, 65–66
   transitional arrangements, 62–63
   World Trade Organization (WTO) obligations, 65
Australia New Zealand Food Standards (ANZ) Code, 60, 63–65
Australian Food Standards Code, 61, 63, 73
Average intake assessment model, 118

B

Beverages, MERCOSUR control, 92–93
Biological markers, sources of, 129–130
Bovine somatotropin (BST), Codex decision making on, 158–159
Brasilia Protocol, 83
Budget Method, 116

C

Carcinogens, Delaney Clause, 32
Center for Science in the Public Interest, 161
Chernobyl accident, 245
Cluster analysis, food supply surveys, 122–125
Code of Ethics for International Trade, 161–162
Code of Federal Regulations, 29
   publication of standards, 30
Codex Alimentarius Commission (CAC), 3, 4–9, 66
   consensus, 162–163
   contaminants and toxins standard, 195–209
   creation of, 4

decision making process of, 155–166
expert opinion, 161–162
exposure assessments, 109–111
on food additives, 5
food additives standard, 171–191
governance of, 164–166
impact of, 149–150
interest group participation, 163–164
international commissions, role of, 214–215
legitimate factors and decisions, 158–159
membership of, 154–155
microbiological criteria for foods, 257–267
objective of, 4, 11
pesticide residue standard, 213–225
precautionary approach of, 159–160
procedural manual, 197
public awareness of, 150–152
public participation at meetings, 152–153
public participation at national consultative committees, 153–154
remit of, 150
risk assessment, 160–161
scientific evidence, use of, 7–8, 157
standards development, 6–7, 139, 196–197
veterinary drug residues standard, 227–233
Codex Alimentarius Commission Procedural Manual, 197
Codex Committee on Food Hygiene, 258
Codex Committee on General Principles (CCGP), 154
Codex Committee on Nutrition and Foods for Special Dietary Uses (CCNFSDU), 151–152, 155
Codex Committee on Pesticide Residues, 122, 215
Codex Food Categorization System (CFCS), 87–89
Codex General Principles of Food Hygiene, 257, 258
Codex General Standard for Contaminants and Toxins in Food. *See* Contaminants and toxins standard
Codex General Standards for Food Additives (GSFA), 89, 108, 113
Codex Principles of Food Import and Export Inspection and Certification, 17
Codex standards
contaminants, classification of, 73
development of, 6–7
European Community (EC) interaction with, 46, 53
and MERCOSUR, 87–89
national/international standards, 9
number of standards, 7
risk analysis, 8

and Sanitary and Phytosanitary Measures (SPS Agreement), 11, 13–24
science, role of, 7–8
and World Trade Organization (WTO), 11–13
Codex Standing Committee on Veterinary Drug Residues in Foods. *See* Veterinary drug residues (Codex)
Color additives
approval process, 33
legal factors, 32
MERCOSUR control, 87
Common Market Council (CMC), MERCOSUR, 81
Common Market Group (CMG), MERCOSUR, 81, 83
Company standards, preparation of, 139
Compliance Policy Guide, 30
of FDA, 30
Confederation of Food and Drink Industries, 72
Consensus, Codex Alimentarius Commission (CAC), 162–163
Consumers
education of, 144
interest groups, 152–153
Consumers International (CI), 152–153
international level, 153
statement on rights of consumers, 152–153
Contaminants
Australia/New Zealand regulation of, 72–76
Codex classification, 73
decision making, criteria for, 89
definition of, 72
FDA regulation, 35–36
MERCOSUR control, 89–91
Contaminants and toxins standard (Codex), 195–209
analytic data criteria, 206
annexes of, 196
chronology of development, 197–204
current status, 208
fair trade considerations, 207
future view, 208–209
intake data criteria, 206–207
maximum limits (MLs), 204, 205
philosophy paper on, 197–198
prevention of contamination, steps in, 205
purpose/scope of, 204–205
risk assessment, 207–208
risk management, 208
technological considerations, 207
toxicological information criteria for, 206
Controlled experiments approach, 121
Cracking the Codex, 155
Critical groups, meaning of, 240
Cultural diets, assessment of, 123–125

D

Dairy products, MERCOSUR control, 92
Data collection
  for dietary exposure assessment, 120–121
  Global Environmental Monitoring System (GEMS), 106
Declaration of Iguaçu, 80
Delaney Clause, 30, 32
De minimus risk, 32
Denner, Dr. W.H.B., 5, 195
Denner Report, 175–176, 195
Developing countries, Codex membership, 155
Dietary exposure assessment
  average intake assessment model, 118
  and biological properties of chemical, 116–117
  Budget Method, 116
  choosing method for, 116
  controlled experiments approach, 121
  dietary intake studies, 121–131
  duplicate diets approach, 120
  intake estimation equation, 116
  joint distribution/Monte Carlo-type probalistic assessment model, 118
  limitations in research, 131–132
  market basket approach, 120–121
  and physical/chemical properties of chemical, 117–118
  purpose/methodology relationship, 116
  quantifying concentration of chemical, 119–120
  range-finding intake assessment, 118
  simple distribution assessment model, 118
  types of data required, 115
Dietary intake studies, 121–131
  basic questions in, 121–122
  chemicals in few foods, 128
  cultural diets, assessment of, 123–125
  food consumption of subpopulations, 128–129
  food supply surveys, 122–126
  food use studies, 126
  household inventories, 126
  individual intake studies, 126–128
  infrequently consumed foods, 128
  intake scenarios, 119
  international estimated daily intake (IEDI), 219–220
  pesticide residues, 219–220
  regulatory commissions associated with, 122–123
  reliability of data, 130–131
  source/consumption location of food, 128
  supervised trials median residue (STMR), 219–220
  theoretical maximum daily intake (TMDI), 219
  validity of data, 129–130
Dietary Supplements Regulations 1985, 59
Direct food additives, 33
  approval process, 33
Dispute Settlement Body, 13
  functioning, examples of, 23–24
Dose constraint, radiation protection, 248
Duplicate diets approach, 120
  method in, 120

E

Egg Products Inspection Act, 30
Emergency exemptions, pesticides, 38–39
Environmental Protection Agency (EPA), pesticide control, 36–39
European Atomic Energy Community (EURATOM), 242–244, 245
European Community (EC)
  fish products, 52
  food additives, 44–46
  heavy metals, 53
  interaction with Codex, 46, 53
  maximum residue levels (MRLs), 46–50
  mycotoxins, 52
  nitrates, 52
  pesticides, 46–48
  preparation/adoption of legislation, 47–48
  scientific advisory committees, 44
  veterinary drug residues, 49–51
European Union Food Control Directive, 7
Expert opinion, Codex Alimentarius Commission (CAC), 161–162
Exposure assessment
  Codex Alimentarius, 109–111
  definition of, 103–104
  exposure consultation, 108
  function at international level, 112
  international assessments, 106–108
  methods, criteria in development of, 112–113
  national assessments, 105–106
Extraneous residue limits, pesticide residue standard (Codex), 222–223

F

FAO/WHO Conference on Food Standards, Chemicals in Foods and Food Trade, 5, 8
FAO/WHO Conferences on Food Additives, 5
FAO/WHO Consultation on Food

Consumption and Exposure Assessment of Chemicals, 106, 123
FAO/WHO Consultation on Guidelines for Predicting Dietary Intake of Pesticide Residues, 122–123
FAO/WHO Expert Committee on Food Additives, 161
FAO/WHO Expert Committee on Nutrition, purpose of, 4
FAO/WHO Expert Consultation on Residues of Veterinary Drugs in Foods, 227
FAO/WHO Food Standards Conference, 5
FAO/WHO Food Standards Programme, 11
Federal Food, Drug and Cosmetic Act (FFDCA), 29–30
   amendments to, 30
   Compliance Policy Guide, 30
Federal Insecticide, Fungicide and Rodenticide Act (FIFRA), 36–37
Federal legislation
   Egg Products Inspection Act, 30
   Federal Food, Drug and Cosmetic Act (FFDCA), 29–30
   Federal Insecticide, Fungicide and Rodenticide Act (FIFRA), 36–37
   Federal Meat Inspection Act (FMIA), 30
   Food and Drug Administration Modernization Act (FDAMA), 30, 32
   Food Quality Protection Act (FQPA), 30
   Nutrition Labeling and Education Act (NLEA), 31
   Poultry Products Inspection Act, 30
Federal Meat Inspection Act (FMIA), 30
Federal Register, 30, 33, 36
Federal standards, preemption and state laws, 29
Federal Trade Commission (FTC), 31
Fish products, European Community (EC) control, 52
Flavoring agents, MERCOSUR control, 87
Flavoring Framework Resolution, 87
Food Act 1981, 59
Food additives
   approval process, 33–35
   Australia/New Zealand regulation of, 66–72
   definition of, 31–32, 67
   direct additives, 33
   European Community (EC) controls, 44–46
   and food categories, 88–89
   functional classification of, 87
   indirect additives, 33–35
   legislation related to, 31–32
   MERCOSUR control, 86–89
   to reduce food-borne illness risk, 35
   safety standards, 32
Food additives standard (Codex), 171–191

   chronology of development (1985–1999), 174–190
   current status of, 190–191
   food category system as basis, 172–173
   future view, 191
   purpose/scope of, 171
   vertical/horizontal standards, 172
Food and Agricultural Organization (FAO), 214
   creation of, 3
Food balance sheets. *See* Food supply surveys
Food-borne illness
   food additive products, 35
   pathogens, 35
   See also Microbiological criteria for foods (Codex)
Food categorization system (CFCS)
   as basis of Codex food additives standard, 172–173
   Codex categories, 87–89
   and use of food additives, 88
Food Commission, MERCOSUR, 84–94
Food consumption studies. *See* Dietary intake studies
Food control systems
   evaluation of, 146
   MERCOSUR, 93–94
   needs related to, 142
Food and Drug Administration (FDA)
   contaminants regulation, 35–36
   food additive approval process, 33–35
   GRAS substances, 35
   Office of Premarket Approval, 33
   pesticides, 36–40
   PMN procedure, 33–34
Food and Drug Administration Modernization Act (FDAMA), 30
   changes related to, 32
Food frequency questionnaire, 127–128
Food handlers
   education of, 144
   microbiological criteria, use of, 267
Food identification, Australia/New Zealand system, 71–72
Food industry
   quality standards, improvement of, 141
   training needs for personnel, 143
Food inspection, needs related to, 142
Food Quality Protection Act (FQPA), 30
   pesticide regulation, 36–37
Food record/dairy method, 127
Food Regulations 1984, 59
Food and Safety Inspection Service, 30
Food Safety and Inspection System, 30
Food standards
   application of standards, 145
   Codex Alimentarius Commission (CAC), 3,

4–9, 196–197
  constraints related to national food
    control, 140
  federal interest in, 141–142
  history of, 3–4
  and legislation updates, 140–141
  levels of, 139
  MERCOSUR, 92–93
  national standard adoption, benefits of,
    145–146
  needs related to, 140–146
Food Standards, Chemicals in Food and Food
  Trade, 137
Food supply surveys, 122–126
  cluster analysis, 122–125
  organizations for preparation of, 122
  purpose of, 122
Food use studies, 126
*Framework for the Assessment and Management of
  Food-Related Health Risks*, 66

G

Gaussian plume model, 249
General Agreement on Tariffs and Trade
  (GATT), 5, 12, 57, 150, 154
*General Conditions to Establish Equivalence of
  Food Control Systems*, 94
Generalized derived limits, radiation
  protection, 248–249
General Principles of Food Hygiene, 259
*General Principles Governing the Use of Food
  Additives*, 4
General Principles for the Use of Food Additives, 69
General Standard for Food Additives (GSFA). *See*
  Food additives standard (Codex)
Genetically modified food, Codex decision
  making, 165–166
Global Environmental Monitoring System
  (GEMS)
  data collection, 106, 120
  Food Contamination Monitoring and
    Assessment Programme (GEMS/Food),
    122, 219
Good agricultural practice, 221
Good Manufacturing Practices
  Australia/New Zealand, 69–70
  FDA approach, 35
GRAS substances, 35
Growth-promoting hormones, maximum
  residue levels (MRLs), 155–156

H

Harmonization, SPS Agreement, 14
Hazard analysis critical control point
  (HACCP), 30

Hazard characterization, definition of, 103n
Hazard identification, definition of, 103n
Heavy metals
  European Community (EC) control, 53
  MERCOSUR control, 90
Histamine, European Community (EC)
  control, 52
Horizontal approach, 46
Hormones, complaints related to SPS
  Agreement, 22–23
Household inventories, 126

I

Imported foods, pesticide regulation, 39
Indirect food additives, 33–35
  approval process, 34
  Threshold of Regulation, 34–35
Individual intake studies, 126–128
  food frequency questionnaire, 127–128
  food record/dairy method, 127
  recall method (24–hour recall), 126–127
Inorganic contaminants, MERCOSUR
  control, 90
Intake estimation equation, 116
Interest groups, Consumers International (CI),
  152–153
International Atomic Energy Agency (IAEA),
  240–242
International Commission on Microbiological
  Specification for Food (ICMSF), 91, 258
International Commission on Radiological
  Protection (ICRP), 238–240
  task groups of, 238–239
International estimated daily intake (IEDI),
  219–220
International exposure assessment, 106–113
  development, historical view, 107
  elements of, 108
  expert consultation, 108
  limitations of, 112
International Organization of Consumer
  Unions. *See* Consumers International (CI)
International Organization for Standardization
  (ISO), 139
International Plant Protection Convention
  (IPPC), 15
International standards
  preparation of, 139
  SPS Agreement, 11, 13–24
  World Trade Organization (WTO), 11–13
International trade
  and Codex MRLs, 224–225
  conference on, 137
  problems related to, 138
Introduction to Radiation Protection, 238

J

Joint distribution/Monte Carlo-type probalistic assessment model, 118
Joint Meeting on Pesticide Residues, 122, 161
Joint Parliamentary Commission, MERCOSUR, 83–84

L

Labeling
  legislation related to, 31
  MERCOSUR control, 93
Laboratory services, needs related to, 142–143
Las Lenas Cronogram, 84
Latin American Free Trade Association (ALALC), 80
Latin American Integration Association (ALADI), 80

M

Mad Cow Disease, precautionary approach to, 159
MAFF Irish Sea Modeling Aid (MIRMAID), 249, 251
Manufacturer's data, pesticide residue evaluation, 218
Market basket approach, 106, 120–121
Maximum limits (MLs)
  contaminants and toxins standard (Codex), 204, 205
  mycotoxins, 90
Maximum permitted concentrations (MPCs), 73–74
  Australia/New Zealand principles, 74–75
  control criteria, 73–74
  establishment of, 75–76
Maximum permitted levels (MPLs), radiation in foods, 246–247
Maximum residue levels (MRLs)
  definition of, 213–214
  European Community (EC) standards, 46–50
  growth-promoting hormones, 155–156
  international problems, 223–225
  international standards, necessity of, 214
  periodic review, 221–222
  pesticide residue standard (Codex), 213–215, 218, 219, 221–225
  veterinary drug residues, 230–231
Measurement error, sources of, 130–131
MERCOSUR. See Southern Common Market (MERCOSUR)
Mercury, European Community (EC) control, 52
Merger Treaty, 242

Microbiological criteria for foods (Codex), 257–267
  chronology of, 258–259
  hazards, categories of, 262
  and intended use of product, 260
  limits, 261
  microbes, types of, 261
  microbiological criteria, types of, 265
  purposes of standard, 259, 264–265
  sampling plans, 262–264
  setting criteria for, 259–261
  situations for risk, categories of, 262–263
  use by food business operator, 267
  use by regulatory authorities, 265–266
Microbiological safety, MERCOSUR, 91
Mineral water, MERCOSUR control, 93
Model Food Act, 58
Monte Carlo-type probalistic assessment model, 118
Mycotoxins
  European Community (EC) control, 52
  MERCOSUR control, 90

N

National consultative committees, Codex, 153–154
National exposure assessment, 105–106
National Food Authority (NFA), 58–59
National Food Authority Act, 58–59
National Health and Nutrition Examination Surveys (NHANES), 119
National Human Exposure Assessment Survey, 120
National Radiological Protection Board (NRPB), 247–249
National standards, preparation of, 139
Nationwide Food Consumption Survey, 119
New Zealand
  food regulation, 59
  See also Australia/New Zealand food regulation
New Zealand Food Regulations, 62–63
Nitrates, European Community (EC) control, 52
Nutrition Labeling and Education Act (NLEA), 31

O

Office International des Epizooties (OIE), 14
Office of Premarket Approval, 33
Organisation for Economic Cooperation and Development (OECD), 122

P

Packaging, MERCOSUR control, 91–92
Periodic review, pesticide residue standard (Codex), 221–222
Pesticide residue standard (Codex), 213–225
  acceptable daily intake (ADI), 217, 219
  acute reference dose, 220
  dietary intake studies, 218–220
  extraneous residue limits, 222–223
  FAO/WHO joint meeting, 217–218
  and good agricultural practice, 221
  international acceptance, problems with, 223–224
  versus international practices, 223–225
  manufacturer's data, 218
  maximum residue levels (MRLs), 213–215, 218, 219, 221–225
  periodic review, 221–222
  pesticide selection process, 215, 217
Pesticides, 36–40
  emergency exemptions, 38–39
  European Community (EC) control, 46–48
  FDA action levels, 39–40
  food additive regulations, 38–39
  imported foods, 39
  international practices, 223–225
  international standards, necessity of, 214
  legislation related to, 36–37
  MERCOSUR control, 91
  unavoidable residues, 38
  U.S. tolerances, 37–38
Pesticides in the Diets of Infants and Children, 36
PMN procedure, 33–34
  exceptions, 34
Poultry Products Inspection Act, 30
Precautionary approach, Codex Alimentarius Commission (CAC), 159–160
Premarket notification (PMN), 32
Produce, MERCOSUR control, 91, 93
Proposition 65, 29, 37
Public awareness, of Codex, 150–152
Public participation, Codex meetings, 152–154

R

Radiation, meaning of, 235
Radiation protection, 235–252
  accident regulation, 245–247
  critical groups, meaning of, 240
  dose coefficients, 237–238
  dose constraint, 248
  generalized derived limits, 248–249
  intake estimation, 250
  international committees, 238–244
  limitations of committees, 244–245

maximum permitted levels (MPLs), 246–247
  principles of, 237–238
  radiation transport models, 249
  radionuclides in foods, dose assessment, 249–250
  United Kingdom, 247–251
Radiation Protection Act 1970, 247
Radioactive contaminants
  effects on humans, 237
  sources of exposure, 236–237
Radioactive Substances Act 1960, 251
Recall method (24-hour recall), 126–127
  limitations of, 127
Redbook, The, 33
Regional standards, preparation of, 139
Regulation of Food Additives, 66
Relative risk, 8
Reliability of data, dietary intake studies, 130–131
Research, needs related to, 144–145
Risk assessment
  Codex Alimentarius Commission (CAC), 8, 160–161
  contaminants and toxins standard (Codex), 207–208
  de minimus risk, 32
  at international level, 105
  international model for, 104
  pesticide residue standard (Codex), 218–220
  professional involved in process, 106
  risk communication, elements of, 104
  steps in, 5, 103
  See also Exposure assessment
Risk characterization, definition of, 103n
Risk management
  contaminants and toxins standard (Codex), 208
  definition of, 208

S

Sampling plans, microbiological criteria for foods (Codex), 262–264
Sanitary and Phytosanitary Measures (SPS Agreement). *See* SPS Agreement
Science
  and Codex standard-making, 7–8, 157
  and SPS Agreement, 15
Scientific advisory committees, European Community (EC), 44
Simple distribution assessment model, 118
Socioeconomic advisory forum, MERCOSUR, 84
Southern Common Market (MERCOSUR)
  administrative office, 84
  adoption of resolution, steps in, 84, 86
  background information, 80

beverages, 92–93
and Codex standards, 87–89
Common Market Council (CMC), 81
Common Market Group (CMG), 81, 83
contaminant regulation, 89–91
creation of, 79
dairy products standards, 92
food additive regulation, 86–89
Food Commission, 84–94
food control systems, 93–94
Joint Parliamentary Commission, 83–84
labeling, 93
microbiological safety, 91
mineral water, 93
objectives of, 80–81
packaging control, 91–92
pesticide regulation, 91
produce standards, 93
size of, 79
socioeconomic advisory forum, 84
structure of, 82
technical assistance, 94–95
Trade Commission, 83
SPS Agreement, 11, 13–24, 150, 257–258
functioning, example of, 21–24
harmonization, 14
negotiation of, 13
provisions of, 14
and science, 15
and TBT Agreement, 18
Standards Code. See TBT Agreement
Standards. See Food standards
State standards, preemption and federal laws, 29
Statistical Office of the European Communities (EUROSTAT), 122
Supervised trials median residue (STMR), 219–220

T

TBT Agreement, 65
scope of, 12
and SPS Agreement, 18
Technical assistance, MERCOSUR, 94–95
Theoretical maximum daily intake (TMDI), calculation of, 219
Threshold of Regulation, indirect food additives, 34–35
Tokyo Round, 12
Total diet studies, 106
Total Diet Study, 119
Toxic Hazards of Certain Pesticides, 215
Trade Commission, MERCOSUR, 83
Trade dispute settlement, World Trade Organization (WTO) process, 18–21
Training, needs of food control personnel, 143–144

Treaty of Ascución, 79, 80, 81
Treaty of Rome, 242

U

U.K. Atmospheric Dispersion Modeling System (UKADMS), 249
United Kingdom, radiation protection board, 247–251
United Nations Conference on Food and Agriculture, 4
United Nations Environment Programme, 240
United Nations Scientific Committee on the Effects of Atomic Radiation (UNSCEAR), 240
Uruguay Round, 5, 12, 57
U.S. Department of Agriculture (USDA)
Economic Research Service, 122
Food Safety and Inspection System, 30
inspection activities, 30

V

Validity of data, dietary intake studies, 129–130
Vegetarians, 129
Veterinary drug, definition of, 227
Veterinary drug residues
definition of, 227
European Community (EC) control, 49–51
MERCOSUR control, 89–90
Veterinary drug residues standard (Codex), 227–233
acceptable residue level, criteria for, 228–229
analytical methods, guidelines for use, 232
Codex approval process, 229
commission relationships in, 229–230
data evaluated, 230–232
maximum residue levels (MRLs), 230–231
purpose of, 227
Vitamin/mineral supplements, Codex assessment, 151–152
Voting, Codex Alimentarius Commission (CAC), 162–163

W

World Health Organization (WHO)
creation of, 3
Food and Agriculture Organization, 138
World Trade Organization (WTO), 11–13, 154
creation of, 5–6
framework of, 12–13
recognition of Codex standards, 196
trade dispute settlement, 18–21

Z

Zero risk, 8